# SOUTH STAFFORDSHIRE

## *Street Names*

# SOUTH STAFFORDSHIRE

## *Street Names*

ANTHONY POULTON-SMITH

AMBERLEY

First published 2009

Amberley Publishing
Cirencester Road, Chalford,
Stroud, Gloucestershire, GL6 8PE

www.amberley-books.com



British Library Cataloguing in Publication Data.
A catalogue record for this book is available from the British Library.

ISBN 978 1 84868 172 9

Typesetting and origination by Amberley Publishing
Printed in Great Britain

# Contents

# Introduction

Place name origins have always been of great interest to the author. In researching them it soon became apparent that the etymologies of the names of towns and villages, which are mostly rooted in the Old English or Scandinavian tongues, differ greatly from the origins of the names of our rivers and hills, themselves more often from the earlier Celtic languages.

When it comes to naming the roads and streets which serve the settlements the origins are rarely the same, save for the ancient lanes which invariably were either destination names or taken from minor place names already in use. From the Middle Ages we find lanes taking the names of adjoining fields and farms, while for the first time individuals throw their family names into the mix. As great areas of farmland were set aside for the rearing of sheep, the manors and great estates provided further inspiration, while by the time the effects of the Industrial Revolution had been felt by the workforce and the population ceased being largely agricultural most of the lanes of England had already been named.

However, when the people moved from the countryside into the towns and cities housing was required on an unprecedented level. Those gentlemen who had funded the building of the factories also paid for the nearby homes to house the workforce and, forsaking all modesty, most often named the roads after themselves and their families. Yet even the larger size of the average family in the nineteenth century would not satisfy the number of names required and other sources were sought. Family names of local dignitaries, figureheads, businessmen, landholders and entrepreneurs were taken and still more were required as the population centres grew exponentially.

After the last world war rebuilding projects proceeded apace and new ideas were put forward to avoid any potentially controversial names. Thus former field names were utilised, having the benefit of being both a traditional local name while also projecting a more rural image than perhaps the place does itself. Of course, fields will normally cover a much greater area than any road or street and thus other roads occupying the same fields were given complimentary rural names. This led to other similar themes, such as fauna, flora, rivers, hills and, when ideas were few, the old favourite of names of other towns, cities, counties and villages was employed, for such a list is virtually unlimited.

In the last thirty years, themes for an estate have become increasingly popular and, as demand for more ideas increases, developers and planners have produced a series of subjects which may well have seemed a reasonable suggestion at the time but the relevance to the area (were there ever one) was soon lost. Thus many of the names are often misunderstood, especially when there are few names to compare. In turn this can lead to later names apparently continuing the theme when that theme was not what it seemed – for example Swan Street and Swallow Lane are more likely to be family names than references to birds.

It should, therefore, be noted that whilst every effort has been made to define all the street names of the areas around Lichfield, Tamworth and Cannock Chase, some names are simply not understood. More often than not this is because the name is a simple one and has more than one potential source, or conversely is so unusual to have no other record. Furthermore there are also a number of names which have been corrupted, often in spelling but also misunderstood at the planning stage, as with the difference between the plural and the possessive 's'.

Within these pages each town and hamlet has been given its own heading. Every road name has been included at the time of going to print, even those where the origin is uncertain. The roads are not listed alphabetically, but grouped together depending upon the origins of their names. Hence all field names are listed together, local dignitaries, historical names, destinations, and modern themes. Where possible the older names come first and the later names follow, although it has not been possible to adhere strictly to this format.

Finally there is a chapter devoted to a few suggestions for the next generation of street names. Obviously the author's interest in history has greatly influenced the subject matter here and surely this is only right, for without the foundations laid down by earlier generations the modern places would simply not exist. These suggestions are simply that and should not be taken as criticism of names already in existence. Each chapter in the history of a town leaves its own mark upon the place in so many ways, and the names of its lanes, roads and streets are a permanent reminder of those earlier times. Let the names chosen in twenty-first century reflect the landmark events of these times, both nationally and locally, while also taking into account earlier people, places and events which have previously not been recognised in such a way.

# Acknowledgements

In researching this book I am indebted to a great number of people who have contributed. Not least are the staff of Lichfield Record Office, who showed infinite patience and dedication in pursuit of their duties. Special mention is also merited for Cheslyn Hay and Great Wyrley Historical Society and in particular Trevor McFarlane, Sandra Alcott, Sian Alcott, Sue Challenger and Kath Perry, who virtually wrote much of those two respective chapters between them. Also acknowledgement is due to Gareth Thomas of Lichfield District Council, Cllr Tony Brookes of Fazeley, Peter Harrison of Hednesford Town Council, Tracy Clarke of Cannock Town Council, and to all those detailed in the bibliography whose research made my task so much easier.

# Chapter One

# Burntwood

Acorn View is a name which can lay claim to the most interesting history in South Staffordshire. The tree in question stood in the middle of what was then Swan Corner, now Swan Island, alongside an old cottage. Following the First World War it was deemed that both cottage and tree had to go as part of the redevelopment, yet the council steamroller failed to dislodge the oak. So Jack Manders was sent for, along with his Belgian shire horse which had had some experience pulling tanks during the war. Firmly chained, the powerful beast made short work of the task. Perhaps further developments around here should recognise both horse and owner too. The name of Swan Island was taken from the Swan public house, here by 1870, and Swanfields was in taken from this later name in turn.

Cort Drive remembers Henry Cort who, having realised that British iron and steel production was quite inferior to that being produced across Europe, designed a steamrolling machine which he patented in 1783. This hammered iron, kept at a steady temperature to beat out the impurites, and later devised the reverberating furnace which greatly improved the quality of the iron produced during the smelting and puddling processes. Just six years after he took out the patent he sold his business and he did not earn great wealth from his invention. By 1820 there were 8,200 Cort furnaces operating in Britain, and by the end of that century the nation was producing more tons of quality smelted iron than the rest of Europe combined.

One of the best ideas for a theme comes in the form of commemorating some of the county of Staffordshire's most famous names. Among those honoured are: the Anson family, Earls of Lichfield whose seat is at nearby Shugborough and Anson Close; Darwin Close recalls Erasmus Darwin (1731-1802), grandfather of the even better known Charles Darwin, who lived in a house near Lichfield Cathedral which is now a museum of his general medical practice, together with his other loves of poetry and botany; and Garrick Rise is named after David Garrick (1717-79), a theatre manager and playwright considered one of the greatest actors ever to tread the boards. Whitehouse Crescent is named after a family long associated with the chase. Jerome Way is named for Jerome Klapka Jerome (1858-1927), born in Walsall, then a part of the county, and employed as a railway clerk, a teacher, an actor and a journalist before producing his best-known work *Three Men In A Boat* in 1889. During the First World

War he drove an ambulance on the western front. His autobiography *My Life and Times* was published a year before his death and contained anecdotes and narratives on a full and certainly diverse life.

The giants of the Industrial Revolution gave their names to a theme headed by Brunel Close. Isambard Kingdom Brunel (1806-59) is remembered for some of the most ambitious and sizable engineering achievements of his time. After working with his father on the Thames Tunnel, he embarked on his own and produced the Clifton Suspension Bridge, Great Western Railway, and three ocean-going steamships SS *Great Western*, SS *Great Britain*, and SS *Great Eastern*.

Others include Boulton Close after Matthew Boulton (1728-1809), who formed a famous partnership with James Watt and established the Soho Manufactory, now a fitting museum to his professional achievements; Newcomen Close marks the achievements of Thomas Newcomen (1663-1729), who built the world's first atmospheric steam engine, in 1712 the first steam engine in the world was used to power a colliery in South Staffordshire. Within a few short years these engines were removing water from mines all over Britain.

Hunslet Road reminds us of the company who produced the locomotives used in the local colliery, there is still an engine of this name operating on the Chasewater Light Railway. Huntsmans Gardens is named after Benjamin Huntsman (1704-76), whose tireless efforts greatly improved steel making techniques in Sheffield. He was motivated by the need for better quality steel which could be used in the production of springs and pendulums in clocks and watches. Macadam Close reminds us of John Loudon McAdam (1756-1836) who changed the way roads were laid to create a hard compacted surface, which was raised and incorporated proper drainage. Metcalfe Close commemorates John Metcalf (1717-1810), who was blind from the age of six having contracted polio. Known to all as Blind Jack of Knaresborough he already had a number of business ventures under his wing when he realised improved roads were urgently required. He was the first man to come forward with a formula for repair and construction of roads that were to last. Hence he constructed almost 200 miles of road across the north of England for which he was paid the princely sum of £65,000. John Metcalf overcame his early problems to live a long and productive life and certainly merits a street being named after him, although it would have been a greater honour if the spelling had been correct.

Telford Close pays tribute to Thomas Telford (1757-1834), who engineered the building of roads, canals and bridges. Trevithick Close recalls Richard Trevithick (1771-1833) who, while a genius as an engineer, did not have the natural ability to sell himself and failed to make the fortune from his inventions that others did. It was Trevithick who built and operated the first steam locomotive on rails, carrying passengers in 1801 and goods three years later. John 'Iron Mad' Wilkinson (1728-1808) made his name from the invention of a cannon-boring machine which produced a spin on the cannonball, making it more accurate and much safer for users. What is devised for warfare is quite often found to be even more useful in peacetime and so it was when the same process was used to make cylinders for the steam engines of Boulton & Watt, and this led to the naming of Wilkinson Close. Wedgwood Close is named after Josiah Wedgwood (1730-95), still the most famous name in pottery.

St Matthew's Hospital was demolished and housing built there. The roads are not named after the individuals directly, but remember the wards which were named in their honour. Thus we find Ashmole Avenue after Elias Ashmole (1617-92), best remembered for founding the Ashmolean Library and Museum at Oxford, who made his name in the legal profession and was made Windsor Herald by Charles II. In 1666 he presented a loving cup to Lichfield, the city where he was educated, records show it was a well spent £23 *8s 6d* for it is now considered priceless. Bodington Close is named after George Bodington M.D. (1799-1882). He was a GP in Birmingham treating tuberculosis using the so-called open-air method, and in 1840 he moved to Sutton Coldfield and opened the sanatorium at the White House. Thereafter he went on to Driffold House and the treatment of the insane.

Cavell Road was named after British nurse Edith Cavell (1865-1915), who remained in German-occupied Brussels during the First World War and helped Allied soldiers escape. Her efforts were discovered and she was executed by firing squad. Clegg Road is named after Clegg House, a social centre for the patients of the former psychiatric hospital. John Clegg was a twentieth-century doctor who served as medical superintendent at the hospital from 1946 and went on to become chairman of the Regional Advisory Committee on Psychiatric Services. Cheshire Close reminds us this was once known as a mental hospital; this road was quite aptly named after the Leonard Cheshire Fund, a charity working to give support to individuals with disabilities in order to provide them with the opportunity of some degree of independence, albeit sometimes only for the briefest of periods.

Freer Drive is named after George Lane Freer (1711-1823), surgeon at the General Hospital from 1793 who wrote many books on medicine and surgery. Gettings Close recalls John Salter Gettings (1847-1928), born in Bilston and who served the Chasetown district's medical needs for over fifty years. His wife was president of the local district nursing association and the nurses' home at Chase Terrace was built in her memory in 1923. Glasscroft Cottages were the cottages owned by St Matthew's Hospital and rented out to the staff, thus keeping them close to hand at all times. The cottages were built on farmland owned by a Mr Jones, hence Jones Lane. Gregory Road took the name of Charles Henry Gregory, who came to Burntwood in 1914 as the clerk at the hospital. Following his retirement in 1948 he was a founder member and first honorary secretary of the Hospital League of Friends.

William Reid served the hospital in three capacities between 1902 and 1946 and gave his name to Reid Close. Upfields was named after the Upfield Cottages built for the staff and taken from an old field name. Selwyn Road was named after the Right Reverend George Augustus Selwyn (1809-78), educated at Eton and then Cambridge he went on to become Bishop of Lichfield from 1867 until his death. Not only was the hospital ward named in his honour but also Selwyn College in Cambridge. Shaftsbury Drive is named after Anthony Ashley Cooper, 7th Earl of Shaftesbury (1801-85), and was introduced to the hospital for his work in fighting for the reforms of the treatment of the insane, improved conditions for the working classes, and to protect the children of the nation.

Sister Dora Avenue was named after Dorothy Wyndlow Pattison (1832-78) who trained as a nurse at the Sisterhood of Good Samaritans. Her devotion to duty and her boundless courage were demonstrated time and again: for example, the smallpox epidemic in Walsall in 1868, then seven years later when typhoid was rife in that town, and in colliery disasters too numerous to count. Her life was tragically cut short when, riddled with cancer, she died at the age of forty-six and is suitably honoured for posterity by the naming of this street.

Nightingale Walk and Nightingale Court are named after the ward which, predictably, remembers Florence Nightingale. However, it also led developers, who had run out of old ward names, to use it as a link to a theme of bird names. This explains the existence of Kestrel Court, Kingfisher Court, Merlin Court, and Osprey Court.

As with many towns, a number of roads are named after other place names. It is almost impossible to know why these places were chosen, indeed they may even refer to (unknown) families whose ancestry hailed from these places and thus took the surname. Examples of such place names, and their meaning, in Burntwood include Blandford Gardens, from 'the ford where blay or gudgeon are found' and pointing towards either Blandford Forum or Blandford St Mary in Dorset; Dewsbury Drive comes from 'the stronghold of a man called Dewi', a West Yorkshire town with the Old English suffix *burh* preceded by an Old Welsh personal name; Fairford Gardens takes the name of a Gloucestershire village, a Saxon place name meaning 'the fair or clear ford'; Henley Close features an Old English place name meaning 'the high or chief woodland clearing', although it is difficult to say where the place comes from as it is one of the most common place names in the country; Stapleford Gardens features a Nottingham place name meaning 'the ford marked by a post or posts'; Wentworth Close is probably a reference to the famous golf course in Surrey, although there are similarly named places in Cambridgeshire and South Yorkshire, all of which come from 'the enclosure of a man called Wintra'; Westbourne Crescent, a place name in West Sussex, is a simple name and refers to the '[place at] the stream'; Yorkshire Way or 'the shire or district of York', the city name itself being 'the estate of a man called Eburos'; and Maidstone Drive which features the name of a town in Kent meaning 'the stone where the maidens gathered', although why they would gather has never been understood.

Of course, there are the destination names too, those named after the places to which they offer access. Boney Hay Road leads to the place known as 'Bondi, Bundi or Bula's island enclosure' where the term island refers to dry ground in land which at least part of the time was distinctly wet, Cannock Road from 'the small hill', Chorley Road takes traffic to what was once 'the woodland clearing of the freemen or peasants', Hammerwich Road is 'the building with the smithy', Rugeley Road takes us to 'the woodland clearing on the ridge of land', Norton Lane has a common place name describing 'the northern tun or farmstead'. And Lichfield Road features a complex place name meaning 'the open land near Letocetum', itself the Roman name for what is now known as the village of Wall and the Latin name speaking of a Celtic place name meaning 'the grey wood'.

Local place names and features have always been a favourite source of inspiration for names, and rightly so. Church Road was originally called Church Lane, it refers to the place of worship built in 1819 and consecrated the following year as Christ Church. Forge Lane remembers the old forge near here on Burntwood Road, with John Rock being the blacksmith in 1899. Gorstey Lea loops around on the farmland known as Gorstey Ley, which has become a regional name. As a place name it is derived from 'the gorsey pastureland', the extra 't' comes from a misunderstanding of local pronunciation. Overton Grange gave us Grange Road. Hillcrest Rise leads exactly where the name suggests. Leigh Avenue also refers to the landscape, although strictly speaking should be 'Ley', meaning a meadow, exactly the same as Meadow View.

Canterbury Road was developed on land once occupied by the vicarage and nearby Church Farm, this reminds us that Canterbury is the home of the Church of England.

A reminder of the rural past of Boney Hay.

Emmanuel Road is named after the landowners and farmers, the Hicks family. It was owned by George William Hicks and left to his brother Emmanuel Hicks on his death – perhaps George William Road did not have the same ring to it. Hospital Road was found as such for the first time just forty years after the Hammerwich Hospital had been opened in Burntwood in 1882. Hudson Drive must surely have come from Job Hudson who owned land all around here in 1860, although there is no written record to confirm this.

The district taking its name from 'the high enclosures' gave rise to all of Highfield Road, Highfield Close, Highfield Avenue and Highfields. Manor Rise is named because in 1717 John Lamb established a warren here as a part of the manor and hence the name. Nether Lane is an old route which did, and still does, lie 'lower, beneath, under' both Green Lane and Padbury Lane. It was named New Road when it was resurfaced in the first half of twentieth century. Overhill Road does indeed refer to a road that crosses from one side of a hill to the other. Ogley Vale is dervived from Ogley Hay Road, itself coming from the landscape. Padbury Lane was held by John of Padbury in 1298, which almost certainly means this family name comes from the place name here. However, there is no record of this name here, which will have told us this was something akin to 'Padda's fortified place'.

Rake Hill Lane was known as Rakehill Lane in the seventeenth century, although the map of 1834 gives it as Stephens Hill before coming back to a form of the earlier name. This isolated name may well be a cartographer's error for the name certainly refers to a rack, not the item of torture but that used for holding the feed for livestock. Springle Styche Lane is first recorded as Springlesuch Lane in 1597, a Middle English name with two relevant elements: *springle*, 'a noose for small game' and *such*, 'the quantity specified'. Such a combination speaks of trapping, the *springle* is a method of attaching a snare to a young branch, the noose is pinned down until the animal trips it and is left dangling in the air (as in so many comedy films). The word *such* is a little more difficult for it could refer to two things, either the amount of tension required when setting the trap or – and this is more likely – a limit on the amount of game which could legally be taken. The Ridgeway follows a ridge of land and virtually names itself, and Springhill Court takes its name from the former Springhill Farm.

Travellers Close takes its name from the Travellers Rest pub, recorded as being here in 1880. Warren Road remembers a huge rabbit warren, recorded as containing 411 burrows with around 3,500 rabbits and covering an area of several hundred acres. Westwoods Hollow's name was chosen by the developers and referred to a family and not one person. The Westwoods first appeared in 1851 as maltsters, farmers and nailmakers and by the end of the First World War they owned four properties in the area, known to the locals as Westwood's Hollow and hence to the name of this road. Wheatcroft Close reminds us of a small field where that crop was grown. Woodhouses Lane is from Woodhouses Road leading to the region found as Wodehousleye in 1374 and referring to 'the home in the woodland'.

More obvious landscape features are seen in Dale Drive; Farewell Drive (not a destination name); Green Lane, which was overgrown with vegetation; Russett Close, which alludes to the colour of the soil; and Redfern Drive, the ferns in autumn. Laurel Drive refers to the tree of that name, as does Lime Grove and Rose Lane.

Coulter Lane takes its name from the farm, which itself comes from a technological advance. During the later Saxon times a culter was first used in Britain, this was the wheeled blade attached to the beam of a plough which makes a vertical cut in the soil. This enables the ploughshare to have a cut start and thus makes it more efficient and easier to use. It is likely the name came here through the name of a farmer, rather than the new plough design, but there is no surviving evidence to prove this. Similarly Fitchett's Bank may well be a reminder of the trapping of the European polecat for its fur, or maybe because of a man who made his living by doing so. Meg View historically has been Megg Lane and, prior to that, Meggs Lane. This would suggest this is a person's name but there is no documented evidence to support this. Morley Road runs past the former Morling Farm, from which the corrupted name comes.

Mount Road was previously known as Ball's Mount and was named after the Blue Ball Inn which was here by 1824, and likely here for fifty or more years before that. Land around here was owned by Elizabeth Ball who, when she died in 1769, left several parcels of land for the benefit of the poor. It may be a coincidence that four acres at Ball's Mount were given over for the benefit of the labouring poor at around this time, yet the dates do match. Nailers Drive reminds us that in Burntwood nailing was an important cottage industry. Peggs Row recalls landowner James Pegg, the development was specifically to house the nurses from St Matthews Hospital and the farm labourers hereabouts. Peter's Lane is tempting to suggest is named after a local called Peter, but there is no candidate for this origin. It is thought to have been named Peter's Lane because, from its first appearance on a map of 1725 to today, the road literally 'peters out'.

Poole's Way is undoubtedly a family name, although there are two potential origins. The favourite, albeit only because they were landowners, were the Poole family who were farmers with a nice little sideline in a bakery and catering around what is now Swan Island. Alternatively we can turn to the vicar of Christ Church from 1852 to 1886 and the Reverend George Poole (1806-89). Prospect Road is used as a term suggesting an outlook – prospects and hope for the future – rather than a view. There can be few more apt names for a road on an industrial estate. Park Road refers to the open land alongside which the road runs, and Park Avenue is derived from that.

Scholars Gate was cut on the grounds of the old Burntwood School. Opened in 1877 for the education of 300 children, it closed in 1989. Sedgemoor Avenue came from Sedgemoor House. Smalley Court recalls Edward Smalley, chairman of the Royal British Legion at Lichfield it was developed by the Housing Association of that body and opened by Margaret Smalley, Captain Smalley's widow, in 1983. Tudor Close and Tudor Park Gardens are derived from Tudor Road, itself after a local farming family. There are three possible gentlemen after whom these places could have been named: in 1834 there is a record of brothers John and William Tudor and in 1880 of one Charles Tudor. Willow Avenue was previously known as Dark Lane and later Black Lane, both perfectly reasonable as it was surfaced with black ash. When the council built homes here they changed the name to Willow Avenue, purely to make it sound more appealing.

A theme of rivers and streams is seen for the following names. In each case the origin or meaning of the river name is given and includes: Anker Close, 'the stream of the hermitage'; Blythe Close, a river name meaning 'the gentle or pleasant one'; Derwent Grove, a river name found four times in England and always referring to 'the river where oak trees grow abundantly'; Dove Close, which reminds us of the river immortalised by the poet Wordsworth, forms much of the border between Staffordshire and Derbyshire and which has a name meaning quite simply 'the dark one'; Hamps Close, a Somerset river name from Old English hamm or 'water meadow'; Leam Drive, the River Leam or 'the river of the elm trees' flows through Warwickshire; Manifold Close – this Staffordshire river has a name meaning 'the winding river' and literally 'many folds'; Mease Avenue, flowing through Staffordshire and Derbyshire and referring to 'the mossy stream'; Penk Drive is a river name which takes its name from 'the hill called Penk', in a process known as back-formation the hill also gave a name to Penkridge; Severn Drive, Britain's longest river, has a name which has never been understood, although the listing of Sabrina 2,000 years ago has been suggested as being that of a pagan river goddess; Stour Close is a river in Dorset meaning 'the strong one'; Tame Avenue is the river which gave its name to Tamworth and speaks of 'the dark one'; Tean Close – this Staffordshire river name means simply 'stream' or maybe 'flowing'; and Trent Close, an ancient river name which refers to it being 'the wandering one' and tells us it is liable to flood.

Saints are collected together in St Benedict's Road – the Benedict of Nursia (480-547) was an Italian saint who founded twelve monasteries; St Luke's Road which is named after the Greek-born author of the third gospel of the New Testament, patron saint of artists, physicians and surgeons and many others; St Mark's Road which remembers the author of the second gospel, patron saint of barristers and Venice; St Matthew's Avenue, after the patron saint of accountants, Italy and Salerno and who died in Ethiopia; St Paul's Road was named after the saint originally known as Saul who was temporarily blinded on the road to Damascus; St Stephen's Road, for the man who is often referred to as the first Christian martyr and who is the patron saint of, amongst many others, Italy, Germany, Belgium, Serbia, Kentucky, deacons, casket makers, masons, horses and headaches; and finally St Giles Road (which apparently did not warrant an apostrophe) which remembers the seventh-century saint who is the patron of an astonishing array of people and places, including Edinburgh, blacksmiths, beggars, epilepsy, forests, hermits, horses, lepers, outcasts, the poor, rams, lorimers (spur makers), sterility, breast feeding, mental illness and noctiphobics (those with an irrational fear of the dark or night).

Browning Road is named after poet Robert Browning (1812-89); Burns Drive remembers Scotland's national poet, Robert 'Rabbie' Burns (1759-96), whose life is celebrated annually on Burns Night (25 January); while Keble Close is after John Keble (1792-1866), a cleric and poet who initiated the Oxford Movement: a review of the sacraments and doctrines of the Church of England.

Street names which give an impression of the countryside are very popular, particularly with developers who feel they make the properties more attractive and thus easier to sell. Names which fall under this category are Bracken Close, Bramble Lane, Elder Lane, Ferndale Close, Foxcroft Close, Foxhill's Close, Glenmore Avenue, Gorseway, Grange Avenue, Keepers Close, Leafenden Avenue, Larkspur Avenue, Moor Close, Mossbank Avenue, Sanderling Rise, Shirelea Close, Squires Gate, Thistledown Avenue, and Woodland Way.

Of course, one of the favourite sources of inspiration is Cannock Chase. Although not all are named directly, there are also those which allude to its location and also satisfy the developers' desire for a 'countryside' name. A franklin was a man with extensive land holdings but not of noble birth, Franklin Drive was chosen to suggest the countryside too, while other examples can be seen in Chase Road, Chaseley Gardens and Deerfold Crescent.

# Chapter Two

# Cannock—Comberford

## Cannock

Dartmouth Avenue is derived from Dartmouth Road, itself remembering when Cannock Chase was made a National Park in 1933, the committee being formed several years before under the presidency of the Earl of Dartmouth. Hatherton Hollow and Hatherton Croft come from Hatherton Road, which itself comes from nobility who helped fund the Pagets.

Hampton Gardens from Hampton Road, Hallcourt Crescent and Hallcourt Close from Hallcourt Lane, and Old Fallow Avenue from Old Fallow Road, are taken from former local place names. Destination names include Cannock Road, Hednesford Street and Hednesford Road, Leacroft Lane, Lichfield Road, Old Hednesford Road, Old Penkridge Road, Stafford Road, Walsall Drive, Rumer Hill Road, with Penkridge Court and Penkridge Mews from Penkridge Way.

One of the oldest and most widely seen sources of street names are place names. The reasons for this are many and the original thinking is rarely obvious and likely to remain unknown. Indeed, it is quite likely the names were chosen simply because they are recognisable as place names. There is a wealth of examples to chose from. In Cannock we find the following: Arran Close, the place name meaning 'the settlement at the peaked hills'; Andover Place or 'the place by the ash trees'; Ascot Drive, meaning 'the eastern cottages'; Ashbourne Close or 'the stream of or by the ash trees'; Banbury Road, 'the stronghold of a man called Banna'; Berwick Drive or 'the barley farm'; Bideford Way, telling us of 'the ford of the stream called 'Byd' and an unknown stream name; Burnham Gardens meaning 'the homestead on or by the stream'; and finally Carlisle Road or 'the place belonging to a man called Luguvalos'.

Chatsworth Drive tells of 'the enclosed settlement of a man called Ceatt'; Cleveland Drive is 'the district of the cliffs or hills'; Clifton Avenue, meaning 'the farmstead by the cliff or bank'; Conway Road, an Anglicised version of the Welsh Conwy, a river name meaning 'full and flowing'; Cumberland Road, a name meaning 'the region of the Cumbrian Britons'; Dorchester Road or 'the Roman town built near the place of the fist-sized pebbles'; Ellesmere Road is 'Elli's lake or pool'; Exeter Road with the Devon city named from 'the Roman town on the River Exe'; Filey Close is 'the promontory shaped liked a sea monster' and a typical Scandinavian name; Girton Road speaks of 'the farmstead on gravelly ground'; Glencoe Drive comes from 'the narrow glaciated valley'; Grasmere Place actually refers to 'the pool called the grass lake'; Haslemere Grove 'the pool where hazels grow'; Kendal Court points to this place being an abbreviated form of 'the place with the church in the valley of the River Kent'.

Kenilworth Court and Kenilworth Drive remind us that this Warwickshire town began life as 'the enclosure of a woman called Cynehild'; Kensington Gardens, 'the farmstead associated with a man called Cynesige'; Kent Mews, assuming it is the county, probably means 'the coastal place'; Kilmorie Road means 'the place with or of the large church'; Kingston Court is another 'king's manor'; Kingswood Avenue is predictably 'the king's wood'; Langdale Drive 'the long valley'; Leamington Close 'the farmstead on the River Leam or elm river'; Matlock Drive has long been an important place in Derbyshire as the name tells us it was once 'the meeting place at the oak trees'.

Meriden Close could be either 'the pleasant valley' or 'the valley of merriment'; Montrose Close is 'the promontory of the peat moss'; Patterdale Road features an old Irish name in 'the valley of a man called Patric'; Portland Place is 'the estate by the harbour'; Repton Close is from 'the hill of the tribe called Hrype'; Salcombe Close 'the salt valley'; Skipton Place is a Yorkshire name meaning 'the sheep farmstead'; Southborne Place is 'the southern stream or spring'; Stirling Place, 'the enclosed land by the stream'.

Strathmore Place takes the name of 'the wide valley' in Scotland; Chalfont Court and also Chalfont Avenue features a Buckinghamshire place name referring to 'the place

Cannock Wood Methodist Church on Chapel Lane.

frequented by calves'; Westbourne Avenue means 'the western spring or stream'; Whitby Way is either 'the white village' or 'the village of a man called Hviti'; and Winchester Road is 'the Roman town called Venta', with the Celtic place name meaning 'the chief place'. A comparison between the accurate description of the original place name with the modern development is often entertaining.

Devon Court and Somerset Place are two names derived from counties. One complements the other, however two names can hardly be considered a theme in the usual sense.

There are always street names which are seemingly place names, indeed probably are; however, there is no recognisable theme and without documented evidence the name may not have come directly from a place but via a family name, a business name, or even coined by a developer for reasons unknown. Hence we cannot be sure of the origins of Alton Grove, Belmont Avenue, Belmont Close, Brooklands Road, Chaseley Avenue, Chaseley Croft, Chenet Way, Cornhill, Canford Place, Dursley Drive, Gresham Road, Grimley Way, Newhall Crescent, Teddesley Court, Trent Road, Ulster Close, Wellington Drive, Wells Close, Smalley Close, Stratford Way, Langdale Gardens and Langdale Drive, Sunfield Road and Waveney Grove.

Camelot Close is off Arthur Street and named to suggest a link with the legendary Knights of the Round Table. However Arthur Street refers to Henry Arthur, a former landowner. Moreton Street takes its name from a nineteenth-century field. Backcrofts is a redevelopment, where there was originally an extensive area of land known as the crofts at the back of the houses. The clayey soil here gave its name to Cley Road. Coal Haulage Lane, of obvious derivation, is here in the middle of the nineteenth century and so is not a comparatively modern name. Common Lane was cut on that field, as was Fallowfield, and Green Lane is a general reference to the arable fields here. Lodge Lane still leads to Lodge Farm under the M6 Toll Road and the A460.

Several streets are named after the churches, including St Aidan's Road, St Chad's Close, and St Lukes. St James Road must also have been an old church name. Short Street certainly lives up to its name, Broad Street less so, but it is certainly wider than many old routes. Hamelin Street, Hanover Place and Brunswick Road could relate to German place names, or the latter two names to German royal households. Haig Close and Lord Cromwell Court must surely refer to two names synonymous with two of the more troubled periods in English history, as Field Marshall Haig was a leading figure in the First World War, while Oliver Cromwell was a leading Parliamentarian during the English Civil War. Foxes Rake features an old landscape name, which indeed refers to the run of the common British mammal, Rokholt Crescent is another landscape name referring to 'the rocky woodland'. Old School Court marks the site of a former place of education.

Addison Close is named after Joseph Addison, an old boy of Lichfield Grammar School and whose father was Dean of Lichfield. Bradford Street comes from the Earls of Bradford, a family associated with the Angleseys. Workers on the Paget estate gave names to streets and include: Harrison Road, Thomas Harrison; Price Street, Mr. D. Price; James Street, Mr James; and Owen Walk, Mr Owen. Samuel Bailey was a mine agent for Lord Hatherton. Boydon Close is named after nineteenth-century landowner Joseph Boyden, Spring Street, after landowner Catherine Spring.

Beecroft Road is an old name in this region, quite likely where hives were found for many years. If so the bees would have been artificially placed in a hibernation chamber

Church of St Luke, Cannock.

or winter box specifically created within the walls of the field here. Brookfield Drive is a reference to an obvious old field name. Brook Vale marks the position of a brook which was certainly here in the nineteenth century. There are three places called Church Street, one off Queens Square refers to St Lukes, a second off Cannock Road refers to Chadsmoor Methodist Church, and the third is near the Watling Street in the district of Bridgtown. Cross Street quite clearly allows people to cross between Moreton Street and Cecil Street without taking a long detour. Delta Way crosses the region between the A5 and A34 which forms the classic shape of a river delta.

The Ridings, Ridings Brook Drive, Ridings Brook, Riverside Court, Millbrook Close, Mill Park and Mill Street refer to the old mill on Ridings Brook, or more correctly the land that surrounded it. Navigation is a more modern example of this same idea. Old Fallow Road is an obvious field name. Park Road and Park Street are always a reference to the grounds of important houses. Lloyd Street is named after Lloyds Bank, here at the end of the nineteenth century.

Danilo Road is named after the old cinema. In the 1890s several people here gave their names to modern streets: Barnard Way remembers Arthur Hart Barnard who was resident at Milford Cottage; Bath Road is named after beer retailer and shopkeeper Edwin Bath; George Green was a butcher in Station Street and gave his name to Green Lane; Hall Meadow comes from Charles Hall, a hardware dealer in Station Street; and farmer Edward Lee gave a name to Lee Walk. Hallcourt Lane, however, is from the old hall which was here before the nineteenth century and which also gave its

A lasting reminder of the old cinema at Cannock.

name to Old Hall Lane. Manor Avenue is located on the site of this building, while Market Hall Street and Market Place are self-explanatory. Hatton Road remembers John Hatton and Hawkins Drive, John Hawkins, both were guardians of Cannock Union Workhouse.

Allport Road and Allport Street take the name of a family recorded to have been here since the eighteenth century, as have the Elliott family of Elliott Close. Burns Street may well be named after Jim Burns, a miner who was killed at work in 1892. Downes Way is named after Alex Downes, a miner crushed between two trucks in 1882. Caxton Court comes from Caxton Street itself, probably commemorating William Caxton, inventor of the printing press. However, Walhouse Street is certainly named after Ann Walhouse who gave two thousands pounds to establish the Walhouse Girls School in 1828. Windrush Road is named from one of the loveliest of the tributaries of the Thames.

Eastern Way is aptly named, as it is the eastern by-pass. East Street, West Street and North Street are all named to link with the others. However, there is no South Street as there was already a South Close and none of them are directionally accurate. Southgate is towards the southern part of the town. Hollies Business Park gave names to Hollies Avenue, Hollies Court, Hollies Park Road and even Hollyoak Way. New Street at Bridgtown is a fairly new development, while the New Street nearer the town centre is approximately a century older. Petersfield must be from a local field name, although there is no record of anyone of that name here. Phoenix Road is a common name for a slum clearance area redeveloped as a trading estate, a reminder of the fabled bird which rises from the ashes, while Progress Drive is similarly a reference to the technical college. Ringway is a part of the modern road system.

Roads named after former members of Cannock Urban Council include Clarke Avenue, Benion Road, Foster Avenue, Rowley Close, Croft Avenue, Hampton Street, Hunter Road and Morris Grove. Bishop Close and Cardinal Way are alongside Cardinal Griffin Roman Catholic School, the inspiration for the name. Cavans Close is opposite Cavans Wood. Datteln Road is named after Cannock's twin town of Datteln in Germany. Elgar Close must be named after composer Edward Elgar, but there is no known link with Cannock. Sherbrook Road takes the name of a local region, as does Shoal Hill Close, Spinney Farm Road, Wellfield Close and Waterbrook Way.

Cemetery Road runs right past the cemetery, The Crescent describes its shape, and Central Avenue describes its location perfectly. Pennine Drive fits with Calving Hill. Field Street leads to the playing fields, while The Firs was named indirectly from the trees which grew here and gave a name first to the cottage and then the road. Gorsey Lane describes the landscape, Leacroft Lane marks an original field name, and Longford Road, Longford Court and Longford Gardens point to an early crossing of water and are part of the original Watling Street, a Roman road which also inspired Roman Court. Watling Street was named after the tribe that inhabited St Albans some 2,000 years ago. Convent Close is off St Johns Road and both refer to the church here.

Railway Street and Railway Walk are adjacent to the railway; Station Court is next to the station. Sandy Lane is still obviously of a sandy appearance. Stoney Lea Road, Stoney Croft and Stoneyfields Close all remind us these were cut on land which was noticeably stoney. Simarly the same idea was applied to Pebble Mill Close and Pebble Mill Drive. Walkmill Lane and Walkmill Way remember the region where the fullers of cloth were located, while Wood Lane is as obvious as it seems. Heath Gap Road is an old landscape name, as is Hickling Road.

Themes are increasingly popular for modern streets and Cannock has several. Royals are featured here in Anne Crescent, named for the Princess Royal; Albert Street, for Prince Albert of Saxe-Coburg-Gotha, consort of Queen Victoria, who is also remembered in Victoria Street; and Edward Street, named after Edward VII. Elizabeth Road speaks of the present Queen and Philip Grove, her husband the Duke of Edinburgh. Margaret Drive is named for the Queen's late sister and Snowdon Way, Margaret's ex-husband. Prince Street is a general name, but Queen Street refers specifically to Victoria. Westminster Road links them all, as it is where all of them were married.

The theme of mythological deities provided Apollo Close, after the Roman equivalent of the Greek Helios; Mercury Road, after the Roman messenger of the gods associated with trade, profit and commerce; Odin Close, after the chief god in Norse mythology, the equivalent of the Saxon Woden and the basis for the name of Wednesday; Orion Way recalls the hunter of Greek mythology; Saturn Road, the Roman god of the harvest and agriculture whose annual feast in late December coincided with the lengthening of daylight hours – this was supplanted by the Christian festival of Christmas – and who provided the basis for the name of Saturday; and Thor Close, a Norse god of thunder who gave a name to Thursday.

Rivers are a popular theme, giving Avon Road, from a Celtic river name meaning 'river'; Dee Grove, a Celtic river name for 'goddess, the holy one'; Derwent Grove, or 'the river of the oak trees'; Don Grove, or 'the river of the god'; and Tame Grove, the River Tame speaking of 'the dark one'. Trees too, in Beech Tree Lane, Birch Avenue,

Cedar Hill Drive, Columbian Drive, Columbian Way, Elms Drive, Hazelwood Grove, Maple Crescent, Oaks Drive, Oakhill Drive, Oakwoods, Poplar Avenue, Redwood Drive, Walnut Drive and Walnut Close, and The Willows.

Lakes of the Lake District gave names to Coniston Way, a name referring to 'the king's manor' and the name of the settlement before it was applied to the lake; Thirlmere Close, and probably 'the place in the hollow by the lake' and again applied to a settlement beforehand; Ullswater Place, 'the lake of a man called Ulfr', an Old Scandinavian name; and Windermere Place or 'the lake of a man called Vinandr'. Carfax, Keble Close, Oriel Close and Trinity Close are all named after Oxford colleges. Melbury, Melchester and Wessex commemorate the novels of Thomas Hardy.

Wordsmiths provided another theme and gave Boswell Road, which is named after James Boswell (1740-95), 9th Laird of Auchinleck, who is best remembered as the biographer of Dr Samuel Johnson; Bunyan Place, John Bunyan (1628-88), a Christian writer and preacher; Byron Place, Lord Byron (1788-1824); Garrick Road, David Garrick (1717-79); Johnson Road, Dr Samuel Johnson (1709-1804), the best known lexicographer; Keats Avenue, romantic poet John Keats (1795-1821); Longfellow Place, American poet Henry Wadsworth Longfellow (1807-82); Masefield Grove, John Edward Masefield (1878-1967), former poet laureate; Milton Road, John Milton (1608-74); Shakespeare Grove, England's most famous writer; Shelley Road, both Mary Shelley and Percy Bysshe Shelley; and finally Wordsworth Close, romantic poet William Wordsworth (1770-1850).

Space is suggested by Gemini Drive, Orbital Way and Voyager Drive. A theme of counties sees Devon Road with Devon Gardens cut off it, Lincoln Drive, Oxford Road and Oxford Gardens, Salop Drive, Somerset Place, Surrey Close, Warwick Close, Worcester Close and York Road. Cambria, Gaelic Road, Scotia and Celtic all refer to Scotland. The region of Hawks Green suggested its own theme in the names of birds. Here we find Capercaillie Drive, Falcon Close, Grouse Way, Hobby Way, Nightjar Way, Nuthatch Close, Partridge Close, Peregrine Way, Pheasant Way, Quail Close, Robins Croft, Sparrowhawk Way, Winchat Drive and Woodpigeon Drive.

A more general theme of plants is seen for the names of Laburnum Avenue, Lilac Avenue, Plantation Road, Spruce Road, Poplar Avenue, Sycamore Road, Hawthorns Road and Pinewood Avenue. Politicians have provided the inspiration for Lansbury Drive, Stanley Road and Bevan Lee Road – the latter referring to former cabinet minister Aneurin Bevan, whose wife Miss Jennie Lee was MP for Cannock at the time. Three of the leading figures in the early days of the Labour Party, Kier Hardy, Robert Smillie and John Sankey are remembered in Hardie Green, Smillie Place and Sankey Road. Speedy Close is named after Mr W. C. Speedy, a former town clerk, and electricity company manager Mr P. Wardle is commemorated by Wardle Close. Redhill Road comes from an old landscape name.

A series of modern names (i.e. since the Victorian era) which, having no true etymology, they cannot be explained as anything other than the kind used by developers in the hope of making the region seem more desirable, be it residential or industrial. Included in this list are Abbotsfield, Adam Court, Albion Place, Anglia Road, Ash Grove, Ashtree Court, Balmoral Drive, Baltic Close, Barnswood Close, Beau Court, Bilberry Bank, Broomhill Close, Buttermere Close, Coppice Court, Curlew Hill, Elmcroft Court, Elmwood Close, Fircroft Close, Sherwood Drive, Fairmount Drive, Fairway, Ferndell Close, Festival Court and Festival Mews, Forrest Avenue, The Glade, Glen Close, Glendale Gardens,

Glendawn Close, Goldthorne Avenue, Grainger Court, Grange Drive, Greenfields, Hall Meadow, Hawkyard Court, Highfield Court, High Gardens, High Gardens Court, Holly Street, Ivy Close, Kingsway, Laburnum Avenue, Larchwood Drive, Lotus Drive, Marshwood Close, Martindale, Meadow Hill Drive, The Meadows, Merlin Close, Moorland Road, Moss Road, Moss Street, Mosswood Street, Mulberry Road, Newhall Gardens, Nirvana Close. Nuthurst Drive, Oak Tree Court, Orchard Avenue, Orchard Gardens, The Poplars, Shirewood, Sycamore Green, Vine Lane, Woodhaven, Woodside Place, while Laburnum Close comes from Laburnum Avenue.

With any attempt to define all the street names anywhere, there will always be those which are difficult to tie down and for a variety of reasons. Cross Street runs off Watling Street at an important junction and this may have been where there was a marker cross or an ancient crossroads. There is a church here, however, and this could also have given this street its name.

Often the hardest streets to define are those named after people, for so many surnames are place names. Furthermore when a family name is involved, that family will often have been around for several generations and have several with that name in each generation. Thus it is often difficult to know which individual provided the inspiration. It is tempting to suggest Bowes Drive is linked to the names of Albert and Edward nearby and is thus a royal name – Elizabeth Bowes-Lyon was in the public eye for many years, being the mother of the present Elizabeth II and wife of the late George VI. Georgian Place is most likely a reference to the former royal house. Princess Street, Queen Street (off Bradford Street), Queen's Square, Queen Street (off Newhall Street) and the Prince of Wales Centre are all further royal references, but which royal they were named after depends on when and why each was named.

Cecil Street has so many potential origins it is almost impossible to decide if this is the family name of Lord Burghley, Marquess of Salisbury, the Marquess of Exeter, or former British statesman Cecil Rhodes. More likely this refers to someone who influenced the local area, but we can be certain it is not named after a learning management system in New Zealand, a computer programming language, a brand of cigarettes sold in Denmark or a 1960s US children's cartoon series featuring Cecil the Seasick Serpent. Vermont Green can only have come from the American state of the same name, but how and why it came to Cannock is unknown.

There is a village in Gloucestershire called Blakeney and while Sir Percy Blakeney is the eponymous hero in the classic play *The Scarlet Pimpernel* it is unlikely that this was the basis for Blakeney Way. However, the origins of Blake Close are obscure, for the name could be taken from innumerable figures both nationally and locally. Hannaford Way probably also has an alternative etymology to two American companies, a Pittsburgh supermarket chain and Samuel Hannaford, architects based in Cincinnati.

Taplow Place is named after Taplow in Buckinghamshire, a place name meaning 'the tumulus of a man called Taeppa' and brought here as a family name. Richmond Close is also likely to come from a family name and ultimately from the place name meaning 'the strong hill'. William Morris Grove may well refer to the man who founded the Morris car manufacturing company and who was made Lord Nuffield.

Lansbury Drive must also refer to a surname, possibly former British politician George Lansbury but unlikely to be actress Angela Lansbury or British suffragette Minnie Lansbury, married to the aforementioned George Lansbury's son, who died aged twenty-three in 1922 after contracting pneumonia following imprisonment for her protests.

Lysander Way must surely refer ultimately to the Greek hero. Lysander, a Spartan general who died in 395BC, was responsible for bringing the long Peloponnesian War against Athens to an end. Clarion Way and Union Street have an unknown history, as is the case for the individuals who gave a name to Adamson Close, Crab Lane, Ellis Walk, Grace Moore Court, John Street, Keeling Drive, Kelvin Drive, Gowland Drive, Holder Drive, Horton Close, Locketts Court, Lovatt Place, McGeough Walk, Muldoon Close, Platt Street, Remington Close, Rigby Drive, Whinyates Rise, Woottons Court and Wrights Avenue.

## Cannock Wood

Baden Powell Close remembers Sir Robert Baden-Powell (1857-1941), founder of the Scout Movement. In 1937 the Marquis of Anglesey released a substantial area of land from his Beaudesert estates for the use by the boy scouts and girl guides and this road was cut on that land. However, Baden-Powell's military career is often overlooked. A lieutenant general in the British Army he served successfully in India and Africa, including seeing action during the Second Boer War and the Siege of Mafeking. It was during this time he started writing his books on military reconnaissance and training, which were read by prospective members of the army while still boys. Baden-Powell is reputed to have maintained the first time he had sex he was not impressed and it resulted in such a headache he never made love again. His wife, and co-founder of the Girl Guides Movement, Olave St Clair Soames' comments on the matter are not recorded. Park Gate Road was the road to the gateway on that estate.

Cannock Chase Colliery Company Number 7 Pit was sunk here in 1868. Hardly surprising to find the locals abbreviated the name to simply Sevens, which eventually officially became Sevens Road. Although there are still plenty of evergreen trees bearing cones and waxy needle-like leaves, the name is not taken directly from them. This refers to the road running past a house called The Pines – hence Pineside Avenue. Little Lane describes this short lane quite perfectly. Chapel Lane refers to the Weselyan Methodist chapel opened here on 7 June 1834.

Other names derived from the locality are Slang Lane, meaning 'a narrow piece of land' and Bradwell Lane, 'the broad well or spring'. Similarly Gilwell Road is either an old place name telling us of 'the narrow spring', or was created to differentiate from the previous 'broad spring'. The trailing plant would have been abundant in Ivy Lane. Uplands Close is aptly named for it was cut close to the highest point of Cannock Chase, as is the road known as High Meadow. Holly Hill Road is where the characteristic evergreen tree with red berries grows, and Sycamore Hill refers to the tree *Acer pseudoplatanus*, a species of maple which bears distinctive, winged seeds. Covert Lane reminds us that this region was used for hunting, being an area where game would find cover in thick underbrush, and Hayfield Hill is a field name, where grass and hay was harvested for fodder. Cumberledge Hill is from the place name 'cumber' which tells of a hindrance on a 'ledge' or shelf of land. This suggests that a track-way existed – albeit overgrown – making progress up this hill difficult, and hence the road gave the name to hill.

## Chase Terrace

Local place names are the traditional source for streets, thus they were not really created but simply taken from existing names. Spinney Close is cut off Spinney Lane, itself forming a straight road with Birch Terrace on the other end; together they mark where a spinney of birch trees once stood. Bells Lane marks the site of the Ring O'Bells pub, and Brackenhill Road speaks for itself, while Bridge Cross Road is a sixteenth-century place name showing how two natural convergence points (bridge and crossroads) have united as a name.

Chapel Street reminds us of the Methodist New Connection Chapel, built in 1883; Chaselands is clearly a reference to Cannock Chase; the sight of Gentleshaw Common gave a name to Common View and Heath View, while the region was known locally as 'The Shaw' and gave a name to Shaw Drive; Croft Gardens is an abbreviation of the nineteenth-century field name of Ryecroft; Eastgate Street led to easterly grazing land; Hill Street is aptly named; and Holly Grove Lane remembers Holly Grove House and the small wood where holly grew. Lodge Rise takes its name from the pub which once stood nearby, Morlings Drive leads to Morling Farm and North Street, somewhat predictably, faces north. Parkhill Road is perfectly named as it runs along the gradient of the hill beside Redwood Park, while Park View provides glimpses of Chase Terrace Park.

School Lane, and from it School Close and School Walk, are roads which come from the collieries of Cannock Chase, albeit indirectly. They were cut to run to the Old Boys' School, opened in 1875 by the colliery for the education of 300 or so sons of the miners' families. The school closed in 1931 due, ironically, to the subsidence caused by long-term mining operations. Bailey Close remembers Samuel Bailey, mine agent to Lord Hatherton.

Ryecroft Drive comes from a field name, Slade Avenue is an old place name coming from *slaed*, meaning 'a small valley'. Radmore Close is a name which can be seen to reflect several moments in the history of this region. First it was a place name, an old name for 'the red moor' and the region behind the Redmoor public house. Next it was applied to the Cistercian Monastery at Radmore from 1140 to 1154, before they moved to Stoneleigh in Warwickshire. Their move was necessary owing to the royal hunting lodge here from 1150, the purported noise from the forestry officials riding past was said to be interrupting their prayers.

The Sportsway leads to the home of Burntwood Rugby Club and the sport of rugby union. Thornfield Crescent could never come from anything but a former field name. Thorpe Street, which itself spawned Thorpe Avenue and Thorpe Close, is probably also an old place name referring to an outlying place, that is a region which was farmed only seasonally. Victory Avenue was erected in 1920 but planned two years earlier following victory in the First World War. Around the same time a few buildings here were served by an unsurfaced road and footpath, which produced a surface which was always permanently puddled and is now known as Water Street. John Wesley founded Methodism and the chapel built here resulted in the name of Wesley Close.

New Street was previously known as Mill Street. Its present name first appears by 1900 and it is thought to have been coined when this area was redeveloped. No Name Road is a delicious name created by the locals. For a long time this road, used by those who lived locally, had no actual name, making reference to it somewhat difficult. As the road gained importance as part of the road network the council officially adopted the popular reference to the lane and commissioned the road signs. Fieldhouse Road takes its name from the property which stood here, itself the home of the local blacksmith who frequented the Red Cow in Wolverhampton, the inn of Thomas Gilpin. The workshop was in Hollow Lane, where he made shell augers for the shipwrights of His Majesty's Docks in London. Apprenticed to him was William Gilpin, son of Thomas, who followed him to London when he moved closer to his major customer. Later William returned to the chase to take over the running of his father's inn and also open his own smithy. It is possible the property known as Fieldhouse originally came from the family who lived there, being named Field, although there are no records to show this.

Destination is a natural subject for a road name. Today the road to a place is still referred to as the something road, even when that is not the official name. Here is Cannock Road, Chorley Road, Ogley Hay Road (Ogley Vale is derived from it), Park Road which leads to Chase Terrace Park, and Rugeley Road.

Other place names are among the most popular subjects for street names. The reasons are many although they probably began as family names, later marking towns and villages of some significance to the individuals and/or the community, and finally simply readily available and accepted street names. This gives us Carlton Crescent, Melford Rise and Melford Grange,

Whether naturally or deliberately, individuals who held power and wealth have been common subjects in the street names of Staffordshire. Angel Croft is named after former councillor and poultry farmer Harold Angell, the spelling adapted at the request of the residents; Attwood Road commemorates Thomas Attwood (1783-1859), a businessman, politician and man of the people. Chawner Close recalls the influential Richard Croft Chawner, who was chairman of both the South Staffordshire Railway Co. and the South Staffordshire Waterworks Co. Then, in 1854, he formed the Cannock Chase Colliery Company, which laid the foundations for the development of the area. The road was not named directly from the man, the clue is found in the adjacent Railway Lane which follows the line of the colliery's track. In 1864 the third of the company's locomotives was named Chawner in honour of the man from Wall.

Jervis Court occupies the site of the Chase Cinema owned by the Jervis family. Miles Jervis opened the theatre in 1925 and his wife took over from him on his death until she retired in 1969 at the age of eighty, twenty-five years before the cinema was demolished. Smith's Close reminds us of the hangars erected by E. F. Smith, the beginnings of an industrial estate and used for storing so much unused material (including ammunition and tyres) it was known locally as The Dump.

Of course the major influence on the region was the Paget family, Marquises of Anglesey who lived at the Beaudesert Estate and thus we have street named Beaudesert. Paget Drive only comes indirectly from this family, a look at the names surrounding this fairly recent development, and named in 1980, will show it was cut off Railway Lane, itself a clue as to its origins. It is named after the tank locomotive Alfred Paget, the second to be owned by the Cannock Chase Colliery Company.

One cannot mention the influence of the Pagets without linking them to coal mining, the main employer in the region for decades as is seen in Miners Way, the name given to the Burntwood Bypass. Benches Close refers to the coal seams on which the estate stands. Known as Benches Fault, it reminds us of a feature of open cast mining, which is to cut flat, bench-like steps in the rock strata in order to reach the coal. Bleak House Drive remembers another open cast mine with a name of obvious derivation, Bleak Moor. Cannel coal, also known as candle coal, burns readily with little residue. It is a valuable source of gas and is the variety of coal which readily lends itself to being carved for ornaments, while also providing the inspiration for Cannel Road. Cinder Road is also named from a coal seam. The Grove and Fair Lady Drive are further reminders of this important industry, the passing of which is still mourned by the locals.

Coulson Close and Mavor Avenue reminds us that the cutting machinery for the extraction of fossil fuels were made by Mavor and Coulson of East Kilbride. The railway line, giving access to the colliery, bisected Chase Terrace. A wooden bridge was built to enable pedestrians to cross between the two sides, and in 1933 it was replaced by a concrete bridge, itself demolished thirty years later. Cross Street marks the place the line was crossed. Eights Croft was developed on the site of the Eights Pit of Cannock Chase Colliery, Fives Grove reminds us that Number Five Pit was known to the locals as simply Five's. Lea Hall Drive is named after the last pit sunk in Staffordshire, in 1960.

Littleton Way was another pit, indeed the steepest seams worked anywhere in Britain were to be found here. Its tunnels of over four miles long saw so many injuries that a ward at Stafford Hospital was named after it. Similarly Park Croft remembers the Park Pit Mine itself from the 'park', a name given to a seam of coal between three and five feet in thickness found right across the county. Another seam gave a name to Robins Road.

Gullick Way is named from the gullick chocks, hydraulic blocks powered by oil and compressed air and used to support the roof in the pits. Griffin Close is named after a locomotive of the colliery, which is also recorded as pulling a train, taking employees on outings to the Beaudesert Estate. The Number Three Pit, known as the Plant, was sunk in 1859 and gave the name of Plant Lane, while the New Plant Inn was built to serve the miners and today Newplant Lane is the modern reminder.

Silverdale Drive is named after the Silverdale Colliery, worked from 1830 to 1998, when it was the last coalfield operating in the area. Two Oaks Avenue is derived from the Fair Oak Colliery Company who sank four mines, all of which had suffered from water seeping in and thus had never reached any depth. The Valley is another road named after a colliery, it was worked from 1874, used as a training establishment from 1947, a mine rescue organisation from 1962, and a museum from 1991.

Wembley Lane is nothing to do with football but is again a reference to coal mining, for this was the location of the new workshop of the Number Three Pit, designed to house every possible piece of equipment required to repair or replace the massive machinery of the mine or of the railways which were vital to take away the coal. With its huge size and impressive skylight it was likened to the most famous building opened in that year of 1923, Wembley Stadium, and hence the name.

Stables Way recalls the wooden stabling erected near here to accomodate the pit ponies for just one week each year during the annual shutdown. For the rest of the year the ponies were housed underground, had their own handlers, even had a number of fields set aside for a variety of food items and bedding. Tales of pit ponies being undernourished are unfounded, at least here, while a life underground did not leave them blind, indeed this area would have provided pleasant enough retirement facilities.

Coal was not all that was mined around here. Much earlier, ironstone, a sedimentary rock also known as siderite, was also mined here. Forges have been here in the forest since the thirteenth century, and ironstone was mined on the Beaudesert Estate and smelted with charcoal, so Ironstone Road is a fitting reminder.

Individuals who gave their name to any place simply because they were synonymous with that area are undoubtedly the most deserving. Their names were applied for their memory rather than their money, position or power. For example Bampton Avenue is named after William Bampton, who was charged with keeping children away from Burntwood Chapel so as not to disturb the worshippers. Joseph Dewsbury Close remembers the baker who produced the loaves in the rear of the shop where they were sold, and also operated a delivery service where the horse and cart bearing his name was a welcome sight.

Robinson Road recalls the Robinson family who came from Willenhall to take advantage of the coal boom, not as miners but as butchers. They had a shop fronting their own slaughterhouse in Princess Street and a complimentary greengrocer's on the opposite side of the road. Clement Robinson also owned property in both Rugeley Road and Spinney Road and was certainly a boon to the town.

Spencer Drive came from the owner of three cottages. Richard Spencer, known to everyone as Dickie, owned the three adjoining homes – he sold number eight, rented out number seven and gave number six to Gladys Bradbury who was to become Dickie's daughter-in-law. These were known locally as Spencer's Row, the name sticking long after the family name had disappeared from the area and quite correctly adopted for the name of the developed road.

However, sometimes these oral histories are all we have and such is the case with Myatt Avenue. In 1896 we know John Myatt was a surveyor here and his descendant, Dickie Myatt, worked as a painter and decorator. Here the unwritten history takes over and it is believed he was contracted to the builders of Ashworth and Stuart, the firm who constructed the new homes here during the 1960s. The Crescent is given to many curved streets, while almost every town's historically most important thoroughfare is known as High Street.

Themes for names not only provide a source but also, as they are grouped together, make them easier to locate. Surely the most common theme in the land is that of trees and, as such, different ways of representing an arboreal idea is always being sought. Here is Alden Hurst, Ashmead Road, Aspen Grove, Beechen Grove, California Grove, Columbian Crescent, Coppice Close, Corsican Close, Lebanon Grove, Lilac Grove, Linden Avenue, Lombardy Grove, Maple Close, May Close, Oregon Gardens and Redwood Drive.

Royals and the nobility have proven the inspiration of Chase Terrace. Balmoral Way is named after Balmoral Castle, the official summer retreat of the reigning monarch it was purchased by Queen Victoria's consort, Prince Albert. Covering over 50,000 acres of Aberdeenshire, it includes significant areas designated as wildlife reserves. An illustration of Balmoral Castle has been featured on the reverse of the Royal Bank of Scotland's £100 note for over twenty years.

Windsor Close is named after Windsor Castle, an official residence of the reigning monarch and the inspiration for the change of the family name from Saxe-Coburg and Gotha to Windsor by George V's royal proclamation in 1917. Windsor Castle is an impressive establishment, not only the largest inhabited castle in the world but also the oldest in continuous occupation. A building begun by William the Conqueror in the latter half of the eleventh century, it has been in a state of almost continuous growth ever since. However, the familiar outline of a medieval fortress is not what it seems, for it did not appear as such until the 1820s and the architectural styling of Sir Jeffry Wyattville.

A hat-trick of references to royal households is completed by Sandringham Close and Sandringham House. This Norfolk estate was purchased by Queen Victoria in 1862, almost a century after it was created, at the request of the then Prince of Wales (later Edward VII). This was in order to provide accomodation for himself and his new bride Alexandra. However, it soon became apparent to Prince Edward that the house and 8,000 acres modelled by Charles Spencer Cowper, a stepson of former prime minister Lord Palmerston, was insufficient for his needs and he engaged A.J. Humbert to entirely rebuild the palace. It is still where the royals traditionally spend the Christmas and New Year period.

While a part of the theme of palaces and residences, Blenheim Palace is not at all a royal residence. Indeed, it has not been linked with the royal family since it was granted to the Dukes of Marlborough by Queen Anne. The inspiration for Blenheim Road in Chase Terrace, the palace was constructed between 1705 and 1724 and was the work of architect Sir John Vanbrugh. The construction, however, ran into financial problems owing to a number of major disputes between the architect and Sarah, Duchess of Marlborough.

The nobility are also represented as part of this general theme. Duke Road features the highest rank of the nobility, a rank derived from the Latin *Dux Bellorum* and conveying a message of 'military commander', as indeed it was when first used to describe the Germanic peoples of the Middle Ages. Earl Drive, named after the rank of the nobility directly below a marquess, the title coming from the Saxon meaning 'chieftain' and also seen in the equivalent Old Scandinavian *jarl*. Below the rank of earl comes viscount and thus Viscount Road, which is first recorded in 1387 and comes from Old French *visconte* and Latin *vicecomitem* and strictly refers to 'the deputy courtier'. Baron Close sees the next rung on the ladder of the nobility, a title from Old French *baron* and Old High German *baro* and describing 'a free man or warrior' which can only be seen as 'nobleman'. As with the royal residences the name of Knight Road does not truly fit with the theme of ranks of the nobility. However, it is a title and perhaps this was the idea behind these names. The etymology of the word takes us to Old English *cniht* meaning 'boy or servant' and was used in the modern sense as early as 1100.

Mowbray Croft is related to the nobility theme, the family being one of the richest and most influential in the nation's history. They were created Dukes of Norfolk and later married into the Howard family. This is the premier dukedom in the land and brings with it the hereditary title of Earl-Marshal and one of the two remaining hereditary peerages allowing admittance to the House of Lords. While Princess Close is certainly named after Princess Street – the latter is erroneous for it was known as Prince's Street until 1915 and had been for at least forty-five years – it does not follow the royal theme and can only be a family name, although written evidence has never been traced.

Famous wordsmiths provided the inspiration for another corner of the town. Byron Close was named after the poet Lord Byron (1788-1824), a leading figure in Romanticism who was famously described as 'mad, bad and dangerous to know' by Lady Caroline Lamb. For a British poet it may seem odd to have him revered as a Greek national hero, yet this is not without good reason for he joined the Greek fight for independence against the Ottoman Empire. It was here that he died, not in battle but from a fever contracted in Messolonghi. Chaucer Drive remembers Geoffrey Chaucer, referred to as the father of English literature. He wrote many works but is best remembered for *The Canterbury Tales,* which, although rarely realised as such, remained unfinished at the time of the author's death in October 1400.

Johnson Road was named after Dr Samuel Johnson (1709-1784), a lexicographer, poet, critic and essayist. He suffered ill health as a child, and, indeed, when he was born his aunt was reputed to have said that, as the infant had not cried at birth and was not expected to live, she would 'not have picked up such a poor creature in the street'.

Kipling Avenue was named for Rudyard Kipling, an author whose unusual first name comes from the village and its reservoir north of Biddulph in Staffordshire. His works were greatly influenced by his time spent in India during the days of the British Empire. Longfellow Road remembers American poet Henry Wadsworth Longfellow (1807-82), whose best-known works include *Paul Revere's Ride* and *The Song of Hiawatha.* Masefield Close reminds us of John Masefield (1878-1967), who was poet laureate from 1930. Scott Way is named after Sir Walter Scott (1771-1832), who combined a successful legal career with his prolific writing.

Shakespeare Road is predictably named after William Shakespeare, who needs no introduction, being probably the most famous writer who has ever lived. Yet, what is not so well known is that he spelled his name several different ways, although it is not understood if this was deliberate or not. Tennyson Avenue is named after Alfred Lord Tennyson (1809-92), regarded as one of the great poets of English literature, while Wordsworth Road remembers William Wordsworth (1770-1850), the best known of the Lakeland Poets.

Other historical figures are remembered by Clive Road, named after Major-General Robert Clive (1725-74), 1st Baron Clive of Plassey, always described as Clive of India; Cobbett Road, William Cobbett (1763-1835), who spent his whole life in pursuit of a fair deal for the farming community; and Mountbatten Close, Lord Mountbatten, 1st Earl Mountbatten of Burma and the last viceroy and governor-general of India.

Not all themes are actually what they seem. Developers have a desire to make their new homes sound as attractive as they can before they have even been seen and achieve this with a selection of rural names. A woodland suggestion has been used for Acorn Close, fittingly cut off Oak Lane and Oaken Gardens.

Ring Road is not a ring road in the traditional sense, one which takes passing traffic around and not through the town in order to reduce congestion, and neither could it be said to be a ring. In fact it is straight with the exception of a small 'dog leg' of around thirty degrees. This is a small-scale arterial road designed to feed the side roads of the Chase Park Industrial Estate.

There are always streets named after places, Cranfield Road, Cumberland Crescent, Deal Avenue, Dunston Drive, Dursley Road, Eastwood Avenue, Filton Avenue, Galway Road, Lulworth Road, Lymington Road, Marton Avenue, Rochester Avenue, Sandown Close, and Stamford Crescent. While we also find those few named for reasons unknown, Albion Way, Ashley Road, Clinton Crescent, Cotton Way, Fernleigh Avenue, Halston Road, Hilton Road, Kingsdown Road, Lambert Drive, Lance Drive, Lorne Street, Squirrels Hollow and Sunnymead Road.

## Chasetown

Place names provided the oldest road names and are still used for modern developments. Indeed, some may be created to appear as if they are early place names. Bank Crescent is on an obvious incline, Barncroft was two words when it was the name of the field, Brooklyn Road comes from Brooklyn Farm, while Brook End and Crane Drive are both probably referring to Crane Brook, even if it is over 200 metres away. Highfields Road comes from a field name, albeit indirectly via Highfields House.

High Streets everywhere are named for being the most important road in the earliest days of a settlement. Hillside Close is from Hill Street, itself aptly named as it is on a steep incline, Spring Terrace here refers to a small spring on the side of the hill. A smaller hill gave its name to Knoll Close. Oatfield Close is a former field name, and Oakdene Road is from Oakdene Farm, while The Orchard was a plantation owned and worked almost single-handedly by Mr Postings.

St Anne's Church gave a name to Church Street, built in 1865 it was the first church in the country to be lit by electricity eighteen years later. St Anne's Close has the same origin. Cottage Close and Cottage Court come from Cottage Lane, which takes its name from the miners' cottages built here. New Gardens is derived from New Street, itself laid out in 1861 but not named 'New' for another twenty years.

The most obvious derivation for a road name is its destination, although it should be remembered this works both ways and may refer to your location as the end of the road. Chase Vale is a clear reference to Cannock Chase and Norton Lane to Norton Canes.

The Paget family were undoubtedly the most influential in the region so it is not surprising to find a number of names reflecting their presence. Anglesey Close recalls they were Marquises of Anglesey, and Uxbridge Court recalls that Henry Paget was created Earl of Uxbridge in 1714.

Other names reflect the mining of the area. Baker Close is derived from Baker Street, thought to be named after Joseph Baker, a coalmaster who bought several plots of land here and personally subsidised the cost of building houses. Brantwood Avenue is a name with a somewhat complex origin. Locally born, John Ruskin was an art critic and social reformer during the Victorian era. He moved to enjoy the scenery of Coniston Water for the last twenty-eight years of his life. His new home, now a museum, was called Brantwood and hence the street name. Union Street was the location of the Union Street Working Men's Club and where the men paid their union fees.

Themes are popular for new roads, with trees probably the most popular of them all. As the name of the road is one of the first things prospective buyers will see and hear, developers like to give names which suggest a more rural location and improve the chances of filling the new homes. Hence we see a plethera of names such as Ash Grove, named after the deciduous ash tree, a tough hardwood, which, at the same time, is very durable and a favourite material. Beech Crescent reminds us of the tree which was originally grown for ornamental purposes. Birch Avenue takes its name from the tree with a name from an ancient language and *bhereg* meaning 'bright, to shine'.

Cedar Road, from which comes Cedar Close, takes the name of the conifer with the aromatic wood most famously used to construct King Solomon's Temple in Jerusalem. Cherry Close should be seen as reflecting the ornamental cherry grown for its blossom rather than the fruit. Hawthorne Crescent refers to the Common Hawthorn and has gained the extra 'e' normally found in the personal name. Juniper Close is named after the tree which produces the sloe berries used to flavour gin, and indeed is a shortened form of *genever* the Dutch word for the tree.

Laburnum Grove features a tree which is probably unnoticed except when it is in bloom and well merits its alternative name of Golden Chain with its cascades of yellow flowers. Lilac Grove remembers the popular blossom, with purple, pink and even white flowers, which are considered by some to bring bad luck if brought indoors. Maple Close is named after the tree whose leaf is featured on the Canadian flag. Pine Grove is not exactly named after a tree but rather a classification of trees. Poplar Avenue is named after the tree grown for paper manufacture, matches, and to produce the boxes in which camembert cheese is sold.

Rowan Grove is named for the tree also known as the mountain ash; Sycamore Road and Sycamore Place remember the trees recognisable by their winged seeds; while Woodland Way is a more general name. Blackthorn Avenue is named after the tree which produces a fruit which can be used for preserves. Copthorne Avenue is not actually named after a tree but a place name, as is the case with Elmhurst Drive, which actually means 'the elm wooded hill', and Hazelmere Drive, meaning 'the pool of the hazel trees'.

Burleigh Croft is an unusual name, for neither of the two possible origins have a particularly strong connection to Chasetown. There is a record of a Mrs Burley living in the general area but she is not known to have done anything to merit the naming of the road. Furthermore, the spelling is different and, although this is not at all unusual, it is unlikely to be the case here where the written record is the only connection. If we accept the spelling there is only one famous individual – Lord Cecil Burleigh, among the most powerful and educated men in the land during the reign of Elizabeth I – with whom this street name could be connected. Yet he has no known connection to the region and thus is seems likely that the name was given by the developer. Burleigh Croft is, therefore, a reminder that the reasoning behind street names was quite often forgotten almost as quickly as they were created.

Edward's Road recalls John Edwards who bought, with other parcels of land, an acre here for the princely sum of £114, which was later divided up for building. Note the name has the possessive 's', which is not only unusual in a street name but in this case is in the wrong place! Elunda Grove was named at the suggestion of Councillor Derrick Brookes after he and his wife had returned from holidaying on the island of Crete. Unfortunately, as with Edward's Road, this is another example of a lack of attention to detail for the name of the village should be spelled Elounda.

King Street marks the bakery business belonging to Joseph King, here in 1880. Station Approach leads to Chasewater Heath's Railway Station. Summerfield Road remembers a local farmer who also delivered his milk to the local community. Wharf Lane allowed access to the canal wharf on Anglesey Sidings. Willett Avenue honours long-serving councillor Harvey Willett.

Davy Close remembers Sir Humphrey Davy, inventor of the Davy Lamp which consisted of a wire sieve covering the naked flame, thus providing illumination but removing the risk of igniting the dreaded damp gas found in mines, which was the constant fear of all miners everywhere. Queens Drive comes from Queen Street, itself previously known as Colliery Road and renamed in 1860 to honour Queen Victoria.

A milestone can be an important stage or turning point in a sequence of events and, although there is a possibility that the name may have been suggested by an ancient milestone said to have been here, the actual origin of Milestone Way is the historical milestone of the opening of Burntwood Bypass.

Paviour's Road is one of those odd occasions when a row of houses gave its name to the road in which they stood. The twenty-one cottages of Paviour's Row were here until 1957, the road being named five years later, and were owned by Johan Paviour (more often given as Pavier), himself resident of Hammerwich Place Farm from 1834 until he died in 1871. Peake's Entry is a narrow alleyway linking Queen Street and High Street and is named after Charlie Peake, or more correctly his butcher's shop and slaughterhouse, which stood at the Queen Street end. Pool Road stands at the end of the eighteenth-century Crane Brook. This was dammed in order to form Norton Pool, a head of water required to feed the Wyrley and Essington Canal, which had become a leisure centre a century later and today is known as Chasewater.

More recent developments are often given names which, on reflection, seem quite arbitrary. In fact there must have been a reason at one time, for someone must have made the decision and yet without a record that reasoning is unknown. Such is the case with: Avon Road, Belvedere Road, Eastcote Crescent, Lawnswood Avenue, Loftus Court, Meadway Street, Newgate Street and Trewern Drive.

## Cheslyn Hay

Local names include Bridge Avenue, which does indeed have a bridge; Chapel Square, which marks the location of a former chapel; Zion Close, where the Mount Zion Chapel once stood; Coppice Close and Coppice Lane, both named after the Coppice Colliery; Cemetery Street, which does run to the cemetery; and Chase Avenue, which reflects the nearby chase. High Street is always the main thoroughfare in the history of a settlement, Low Street is named in comparison and could even be said to show the two one-way systems in the modern road system, and Cross Street allows passage between the two. Landywood is an old place name meaning 'the woodland pasture' and gave a name to Landywood Green and Landywood Lane. Lodge View is named after Lodge Farm. Brook Road marks the position of a brook which may well still run underground. Pool Meadow is an old field name. Queen Street commemorates the reign of Queen Victoria.

Campians Avenue takes its name from nearby Campians Wood. Coltsfoot View was inspired by the coltsfoot flowers which grew wild here. Gilpins Croft took the name of a family who owned a local engineering company. Grassmere Court was developed on the land owned by Arnold and Leslie Hawkins, their home being called Grassmere. Glenthorne House was the home of Thomas Hawkins, owner of T. A. Hawkins & Son colliery. Harrison Close was named after the family of mine owners and Collier Close after the workers. Hospital Lane was the site of the former isolation hospital. Leveson Avenue was land owned by a family of that name, Sutherland Road takes its name from the Duke of Sutherland. The Nook is named after the Nook Colliery, which closed in the 1940s. Wesley Avenue is named after William Wesley, a man who served as a parish councillor and was also the treasurer and secretary for Cheslyn Hay Miners' Union. Hatherton Street is also a mining family name. Quarry Close is a reference to the colliery. Charles Avenue remembers landowner Thomas Charles.

Mitre Road takes the name of the Mitre Fault, an ancient geological fault underneath the town. Old Horse Road was an old track which, when resurfaced for modern traffic, was renamed New Horse Road. Off this is cut Sandown Road, which perpetuates the idea of horses. Tennscore Avenue is near the Ten Score Pool, a popular target for anglers, and Cheslyn Drive one of the few named after the town itself. Dennfield Drive refers to the Denny family, market gardeners here in the 1920s and 1930s, who sold off the land for development at the end of the twentieth century. Dundalk Lane is traditionally held to be the homes of navvies from Ireland who were building the Wyrley and Essington Canal. Within living memory it was known locally as 'Dunduck Lane', which may suggest the Irish pronunciation of Dundalk, from where the workers came.

Pinfold Lane marks the position of the pinfold, an enclosure where stray animals, rounded up by the pinner, were held until they were claimed by their owners upon payment of the requisite fine. The money paid for the pinner, the upkeep of the pinfold and prevented stray animals from wandering around the settlement. Highfields Grange is cut off Highfields Park, both named from the 'high fields' here previously. Similarly, Hill Street is on a noticeable slope.

Saredon Road is a destination name as is Wolverhampton Road, Short Street is aptly named for it covers very little distance and Station Street did not exist before the railway arrived. Watling Street is derived from the Waecling tribe who lived on this road near modern St Albans.

Themes found in Cheslyn Hay include that of birds, and the street names Eagle Close, Falcon Close, Hawks Close, Kestrel Way, Lapwing Close, Raven Close and Swan Close can be found here. The theme of the countryside is also prevalent, including The Croft and Forest Glade. Herbs provided the inspiration for Fennel Close, Rosemary Avenue and Rosemary Road.

However, the prize for the most original theme goes to that of the Tudor dynasty, and in particular Henry VIII. Cheslyn Hay features Tudor Way and Tudor Close, the family name of the house that ruled England from 1485 to 1603.

Other references here to Henry VIII rely more on the king's most famous trait, his many marriages. Boleyn Close does not remember Mary Boleyn, long-time mistress of the king, but her sister Anne Boleyn who was to become his second wife and the mother of the future Queen Elizabeth I. Their marriage ended when she was found guilty of high treason, adultery and incest and executed on 19 May 1536 in the Tower of London.

Seymour Close reminds us of the third wife, Jane Seymour, who gave birth to the king's only male heir, the future Edward VI. Reputed to have been the one wife whom he was truly devoted to, Jane died after giving birth to Edward. Cleves Crescent is named after Anne of Cleves, Henry's fourth wife. German-born Anne was married in January 1540 and the marriage, which was never consummated, was annulled in July the same year.

Cranmer Close recalls Thomas Cranmer, the man who engineered Henry's divorce from Catherine of Aragon on the grounds of her previous marriage to his late brother Arthur, Prince of Wales. He was also a leading figure in the English Reformation and served as Archbishop of Canterbury during the reigns of Henry VIII and Edward VI. He was burned at the stake on the orders of Queen Mary I and is regarded as a martyr.

Mary Rose Close is named after the finest warship in Henry VIII's naval fleet, which sank unexpectedly off Portsmouth in 1545. It was famously raised from the seabed in 1982 and is now on display in Portsmouth. What is not generally known is the *Mary Rose* had already served the nation for thirty years, and was carrying 91 guns, 200 sailors, 185 soldiers and 30 gunners when she sank.

The last road relevant here is Anstree Close, which, while cut on the same estate, appears to have no connection with the Tudors, or indeed to anything, for the only reference to Anstree anywhere is in this road name.

Chestnut Drive owes its name as much to the edible nut as to the tree. Hayes View Drive gives a view of Littlewood, exactly as the name suggests, while Hawthorne Road is one of several others alluding to the woodland, as do Littlewood Lane and Littlewood Road, and also Wood Green, Woodland Drive and Woodland Lane. Oakridge Drive, Valley Green and Woody Bank give a theme of the countryside. Park Close is cut off Park Street, both opposite the recreation ground and named from it.

Mount Close is derived from Mount Pleasant, a name early developments were fond of for much the same reasons as modern roads are named to convey a rural feeling. For example although Peace Close is not a field name it is likely to have been created to appear as such. Besom Way, Chiltern Close, Clover Ridge, Old Falls Close, Orchard Close, Oakdene Close, Hazelwood Close, Robins Close, Rosewood Park, Spring Meadow, Thornley Croft and Westbourne Avenue are all recent developments named for similar reasons.

Berwyn Grove takes the name of a place in Wales, Bramwell appears to be a family name, Frensham Close is a place in Surrey. Magna Close is no larger than others, despite what the name suggests, and March Close, Merrill Close, Moon Lane, Wallace Court were all named for reasons unknown.

## Chorley

After Dodds Place Farm, Dodds Lane is probably named after a person named Dodd, but there is no written proof. Shute Hill is a name which has been here since Shute Hill Farm and before. It refers to a slippery slope, manually cleared to give a quick and easy method of delivery for timber cut at the top to be sent to the bottom – in effect a chute. A tithe was a tenth part, paid as a tax in the form of first produce and eventually cash. There was a tithe barn in Tithe Barn Lane in 1882 where the goods would have been deposited.

Lodge Lane is indeed where the lodge was built, and Summerhouse Lane was once part of the estate of ancient Chorley Hall and this path led to its summerhouse. Lower Lane roughly follows the course of the Maple Brook, in the natural dip below Padbury Lane and Green Lane, while Ford Lane crosses it. Little Pipe Lane is a destination name, as Little Pipe and Great Pipe were part of Lichfield's parish. The name refers to the hill which, in turn, took its name from the tiny brook which ran in a particularly narrow and deep channel it had cut for itself – literally 'piped'. Green Hill was blanketed by vegetation.

## Clifton Campville

Smithy Lane remembers the forge of Thomas Roberts, which was here in 1896 and was certainly worked prior to this. Coppice Lane is aptly named for it runs past Ash Coppice, Spring Coppice and Coppice Well, minor place names and which must have existed before the road itself. Chestnut Lane is a very old name clearly referring to the trees which once predominated here much more so than they do today. A new

The site of the old ford at Chorley.

development is named after the church in St Andrew's Close, with St David's Drive named to complement the former, as they are the patron saints of Scotland and Wales respectively. Tudor Rise has retained the name given to it by the developer.

Main Street is still the major thoroughfare in the village, while the Derbyshire villages with names meaning 'the farmstead associated with a man called Lulla' and 'the lower place of the small wood' are reached via Lullington Road and Netherseal Lane respectively.

## Comberford

Elford Road leads to the next village north of Tamworth and is part of the main road between Burton and Tamworth. Tollgate Lane shows the point on this main road where there was once a fee payable for travellers. Furthermore, there is even evidence of who the toll was paid to for this in turn leads to Manor Lane, which is self-explanatory. The toll paid for the upkeep of a road which would otherwise have become so worn and rutted to have been useless.

Payment is no longer collected on the road at Comberford.

# Chapter Three

# Drayton Bassett

Drayton Lane is the original place name speaking of 'the portage by a farmstead'. Portage, now a largely obsolete term, refers to where boats and cargo were dragged across land. Often this was to avoid obstructions in a river, such as a weir or even where the river was simply too narrow or shallow to carry the boat. Similarly Portleys Lane refers to 'the portage by the meadows'. Salts Lane was the original route into the village, it could well have been a part of the so-called 'Wendley Way' (taking its name from Windley in Sutton Coldfield) an ancient salters' route linking the salt town of Droitwich Spa in Worcestershire with Salt in Staffordshire, via Birmingham, Tamworth and Drayton Bassett.

Eddens Wood is still visible in the background at Drayton Bassett.

The moat in question at Moat Drive is the standard defensive feature constructed around the original settlement at Drayton, protecting the old manor house. It is also seen in Old Manor Close, the site of which extends under the churchyard. Eddens Wood Close is after Eddens Wood to which it leads, itself clearly derived from a personal name. Heathley Lane features a field name meaning 'the heath by the woodland clearing'.

Peel Close recalls the family whose name is still synonymous with Drayton Bassett. Indeed, it is more than a surprise to find the village, which virtually provided living accommodation for those who worked on the estate, has only this sole mention of the family. In fact, there are more mentions of the most famous member of the family, Sir Robert Peel, MP for Tamworth and founder of the modern police force, elsewhere in Tamworth and its surrounding districts. Aside from the streets we find the Sir Robert Peel Hospital at Mile Oak and the Bow Street Runner, a bar in the Castle Hotel in Tamworth and a reminder of the early police force.

Church Close is derived from the parish church of St Peter's. Rectory Close is not the normal building one would associate with the name but a large mix of buildings from many eras. It once included stables and outbuildings, and it would have housed rectors and an impressive retinue of servants. When built it was certainly new, but that has not been true of New Row for a hundred years.

# Chapter Four

# Eldingale—Elmshurst

## Edingale

Edingale is mentioned in the Domesday Book, albeit then listed under the county of Derbyshire of which it was then a part. Indeed, even today, one can almost throw the proverbial stone into the neighbouring county. As the subject matter is that of place name origins, it is interesting to note that the pronunciation of this place has changed and possibly within living memory. Old records show the first element of this name to be from the personal name Eadin, pronounced as in 'heed', not as in 'head'. If the latter was the case, throughout history the name would no doubt have evolved to Eddingale and, while the modern form has only a single 'd', it is pronounced as if it was 'dd'.

Arthur Rowley served no less than fifty-three years as a parish councillor for Edingale, was also a school governor, a member of the village hall committee, a founder of the football club, a warden of the church and a trustee of the Maria Booth Charity. His efforts were rewarded during his lifetime by an invitation to a Buckingham Palace garden party and Rowley Close is a fitting tribute to a man whose life was the village.

Holland Court remembers Edward Joscelyn Holland, MBE, who came to Edingale and brought fame through his skill with shire horses. The Hatchett family were landholders here and in neighbouring Croxall since at least 1586; Hatchett Lane was chosen for they provided many churchwardens. They had, however, all moved away from the area before the end of the nineteenth century. Frank Blakeway and his wife Marian, both Edingale-born, moved into The Firs when the Strattons sold their holdings in the village in 1901, Frank being a nephew of the family. Blakeways Close was developed on farmland that was part of that held by The Firs. A directory of 1872 lists one John Moore as a brick and tile maker, butcher and a baker, and is commemorated by Moores Croft. Schofield Lane is after landowner Thomas Schofield (1740-1829).

## Elford

Destination names in Elford are Tamworth Road and Burton Road, effectively the same road leading in opposite directions away from the village, while The Square is a common enough sight in many villages.

While Croft Close and Victoria Meadows are old field names, Peggs Lane seems likely to refer to someone who was formerly in residence here, although no records exist to prove this. Brickhouse Lane marks the place where the first brick houses were built and also, possibly but not necessarily, where the bricks themselves where produced.

Near the church are a number of roads which remember the hall here, indeed there is a road named Old Hall Drive. The Shrubbery and The Gardens are also roads, quite clearly cut on those regions of the grounds. The Beck does refer to a small stream, but this is a very small tributary of the Tame. The watery theme is often attributed to Eddies Lane. It is, however, more likely to have been a reference to an old resident, although that person's identity is a mystery.

Church Road provides access to St Peter's Church, a place of worship said to have been visited by Henry VII and earlier still by Robin Hood. Inside, those who inhabited Elford Hall still leave their mark, but they are over-shadowed by a young boy. John Stanley was an otherwise unremarkable individual, he had not lived long

Nothing 'square' about this street in Elford.

A village which takes pride in its appearance is recognised.

enough to make his mark and we are not exactly sure when he died, although it was probably in 1460. However, we do know how he died, for the church shows him holding a tennis ball in his left hand and his right held against his temple, which was where the ball hit and killed him as he was playing. Perhaps John Stanley also merits a street named in his honour.

Commemorating Sir George Powell Bowles of Elmshurt Hall, in St Chad's, Lichfield.

## Elmhurst

All names here refer to the history of the hamlet. Park Road refers to the estate of Elmhurst Hall, and Nash Lane to Thomas Nash who was resident here in the late nineteenth century. Fox Lane is derived from an old field name. Fox Edge may have existed before the field name and thus refers to an even earlier place name, now lost. Tewnals Lane, listed as Tunals Lane in the late eighteenth century, and a Tunals Lane Farm still exist. The modern spelling is clearly a corruption, but in either case the lack of early forms makes the origins unknown.

Less than two centuries ago the population of Elmhurst was under fifty, yet by 1861 the census showed a total of fifty-seven most of which were farm labourers. However, not everyone worked the fields; the survey records there were two people making ropes, a schoolmistress, a platelayer, a wheelwright and a maker of clock cases.

# Chapter Five

# Farewell—Fazeley

## Farewell

Farewell Hall Mews is from the place name meaning 'the clear spring'. Stoneywell Lane is a farm name, meaning the 'stoney land by the well'. Cross in Hand Lane features the name given to a tiny settlement in the latter half of the seventeenth century, its name coming from 'the cross with the hand' (a fingerpost) here in 1675 but its absence is noted in 1728. As a street name it is first mentioned in 1882.

The construction of St Bartholomew's Church comes from a number of different eras. Originally belonging to a priory of Benedictine Nuns it was founded in 1140, it was repaired and extended in 1527, 1747, 1848 and 1920.

Delightful seat in the grounds of Farewell's church.

St Bartholomew's Church at Farewell.

## Fazeley

Undoubtedly the major influence on Fazeley over the last two centuries has been the Peel family, both directly and indirectly and in a number of ways. While Peel Court takes the family name, Drayton Manor Drive served their estate of that name, Swiss Lodge Drive was named for the style of building found there. Victoria Road and Albert Steet recall the visit to the Peel estate in 1843 by the queen and her consort.

Destination names are also represented by Atherstone Street, Coleshill Road and Coleshill Street, Lichfield Street and Tamworth Road.

Brook End marks the termination of the Bourne Brook, and of course Bourne Avenue refers to the same thing. Broomfield Avenue remembers the club of that name, itself an old field name. Mill Lane led to the old mill, and New Mill Lane, to the more recent mill on Bourne Brook. New Street was aptly named when it was first developed, and Tame Court takes the name of the nearby river. Marina View is a new development which affords excellent views of the marina. The Anson family were still resident at Bonehill Lodge at the end of the nineteenth century, hence the name of Anson Court.

Tolson Avenue remembers one of Fazeley's major employers. Wm Tolson Ltd was the mill known to all as Tolson's Old Mill and this road was cut on this land. William Tolson took over the mill from the Peels in the 1830s and, fifty years later, had built his own factory. The huge building on the banks of the canal was, as with the vast majority of Victorian factories, powered by steam. The adjacent canal and the springs not only supplied ample water to drive the machinery, but also provided the perfect artery for transporting goods into Birmingham and beyond.

Buxton Avenue marks the family who include Samuel Buxton, owner of the Bleachworks in 1835, and Dr Thomas Buxton, landowner in 1939. Mayfair Drive marks the location of the Mayfair Garage, demolished in the late twentieth century.

The Deer Park Estate has its own theme reflecting the deer found on the former Drayton estate. Deer Park Road, Fallow Road and Reindeer Road speak for themselves, yet others are much less obvious. Sambra Road features the name of a deer native to southern Asia, Rangifer Road is named after the scientific name for the caribou; and Dama Road, the scientific name for the fallow deer. Finally there is Mayama Road, which should have been named Mazama after the deer from South America. However, it seems as if someone has written it longhand and it has been misread as a 'y' instead of 'z'.

As with most places there are a number of names which have been coined for reasons unknown and, without records being kept, the origins are a mystery. Beekes Croft, Burlington Court, Tongue Terrace, Victory Terrace, Wrighton Grove and Yorksand have produced a number of suggestions, all of which appear to have been given after the naming of the places rather than the reverse.

The site of the former garage.

# Chapter Six

# Gentleshaw—Great Wyrley

## Gentleshaw

Watery Lane runs down from a natural spring and is virtually permanently wet, Commonside literally runs alongside Gentleshaw Common and Green Lane was blanketed by vegetation. Redmoor Road comes from Redmoor Brook, appearing red because of the high iron content. And Maple Brook Lane, named after the Maple Brook, a tributary of the Redmoor, is named from the trees along its banks.

Windmill Lane, named after the old Gentleshaw Windmill, also gave us Windmill Close. The remains of the windmill are still standing. Built in the early nineteenth century, it was operational by 1818. However, it had a comparatively short lifespan and is documented as being abandoned by 1902. It also gave a name to the Ye Olde Windmill public house, and it stands in the grounds of that establishment. Hence, the age of the windmill suggested by the name of the pub is something of an exaggeration.

Chapel Lane recalls Gentleshaw Chapel, built here in 1830, Gentleshaw School was opened in November 1878 on School Lane to educate 100 children, and Malthouse Road was where malt was soaked, roasted in a kiln to force germination and then used in brewing – the brewing house in question is dated 1763.

Thornleys Hill may appear to be a place name and that final 's' suggests it may be possessive and refer to a farming family.

## Great Wyrley

Minor place names have given Brook Lane and Brooklands Avenue, from Brook Farm and ultimately referring to Wyrley Brook and Maple Hayes Brook. Gorsey Lane and Holly Lane are a traditional names from the local fauna. Wharwell Lane speaks of leading to 'the spring where hares are seen'. Estridge Lane is recorded as the field name of Hestridge in the nineteenth century, meaning 'the east ridge'. Shaws Lane is probably an old place name referring to 'the small wood' but it is just possible it may have come from an unrecorded family name. Moat Farm and its Moat House gave the name of Moat Lane. Pool View overlooks the pool alongside the Watling Street. St Marks Close leads to the church dedicated to this saint. Station Road was developed at the same time as the railway station. While a visit to Tower View Road makes the reason obvious for both the water tower and post office tower are impossible to miss.

There is also Landywood Lane leading to 'the pasture in the wood', while Upper Landywood Lane differentiates from the other by stating it is higher up. Barncroft was developed on the land at Whitehouse Farm where the barn stood, and Paddock Lane passed the fields of this farm. Broadmeadow Lane is cut on the field known as the Broad Meadow. Foxlands Avenue does indeed refer to the hunting of foxes. Fairoaks Drive has taken the name of a nearby coalmine. Leacroft Lane led to the Leacroft Pit, Plants Close was where twenty houses were built by the Plant family to house the miners. Neddy Quintons Colliery was renowned as one of the dirtiest locally, indeed it was said that a 'Neddy' could be spotted at a distance simply from the colour of his shirt; Quinton Avenue commemorates this family.

The lane initially used by miners to access the Grove Pit (also known as The Sinking) was known to the locals as 'Slacky Lane', a reference to the coal dust that fell from the workers' boots as they trudged home. As it was also where the homes of the mine managers were sited it was given the more dignified official name of Hazel Lane. Similarly, Hilton Lane was known as 'Watery Lane' for its marshy location, and Hilton Lane was chosen although it seems to have little etymological basis. Hall Lane and Park Lane remind us of the area where a landscaped greensward was overlooked by Brook House and Old Wyrley Hall.

While locally, landholder John Benton, here in 1792, gave his name to Bentons Lane and his home of Jacob's Hall to Jacob's Hall Lane. The same record shows John Smith and Thomas Jones as resident here, hence the names of Johns Lane and Jones's Lane. Kingswood Drive does still afford distant views of the Kingswood beyond Watling Street. Manor Lane is from Manor Farm, Moat Lane is from its former moat and Well Lane and Wharwell Lane remind us of the longtime spring here and its traditional associations with the English Civil War. Cromwell Court reminds us that members of that family, descendants of Oliver Cromwell, were resident in Staffordshire. Cadman's Lane reminds us of Joe Cadman, former parish clerk. Julian Close was built by Ken Elwell, the road being christened around the same time as his new son and with the same name. Similarly Wheatley Grove is named from the Wheatley Brothers, builders.

Pendrel Close remembers the family of Colonel Leveson. As Royalist supporters they were among the first to receive a demand from their new Parliamentarian landlords for payment of taxes. As owners of Boscobel House they had played host to King Charles during his flight from the Battle of Worcester, hiding in the famed Boscobel Oak. Dove Hollow is named after a man by the name of Plant who used to keep doves and resided in the old houses which made way for the modern road.

Destination names include Norton Lane to Norton Canes and Walsall Road. There are also names taken from other place names, for nobody would expect Cotswold Avenue, Oxford Close, or Telford Avenue to lead to those particular destinations. New Street was, as with all roads of this name, once 'new'. The name of Love Lane developed from the local name for the place, a favoured haunt of courting couples. Lichfield Close is taken from the nearby city.

The Crescent always describes the shape of the road. Sunbeam Drive and Sunset Close were developed and named to complement one another, referring to the first last rays of the rising and setting sun as they face east and west respectively. Dark Lane was poorly lit, as it was so overgrown. Holly Lane comes from the field name of Holly Leasowe. Similarly Field Lane, Lea Lane, and Meadow Grove refer to the earlier field, lea and meadow without taking the names which they undoubtedly were known by even if unofficially. Wyrley Old Hall gave its name to Olde Hall Lane.

Country estates have provided the names for one theme. Somerford Close is named after the eighteenth century Somerton Hall at Brewood, a listed building which now serves as a conference centre and can be hired for weddings and similar functions. Chillington Close marks Chillington Hall, also at Brewood, the home of the Giffard family. It is a Grade I listed building designed by Francis Smith in the early eighteenth century and completed by John Sloane some fifty years later. The grounds and lake were landscaped by Capability Brown, renowned for his talent. Weston Drive is named after Weston House and Park, and 1,000 acres of parkland, laid out by Capability Brown, surrounds the present seventeenth-century house, home to the present Duke of Bradford. Dunston Close marks Dunston House, an eighteenth-century red brick building with additions from the nineteenth century.

However, Pendrill Close is a little different, for it does not mark the house itself but workers on the Boscobel estate, owned by the Giffard family. There were five Pendrill brothers working on the estate: George, Humphrey, John, Richard and William. They were all to play significant roles in the most famous episode in the history of the place. Following the Royalist defeat at the Battle of Worcester, where a sixth Pendrill brother was killed, Charles II, Colonel Carless and Charles Giffard fled to nearby Whiteladies. Here, Richard Pendrill disguised the king as a woodman – even cutting his hair and staining his face and arms. An abortive attempt to escape to France through Wales saw them return to hide in the famed Boscobel Oak. The brothers Pendrill dispatched to arrange the eventual escape of the king on Humphrey Pendrill's mill horse through Bristol, the five brothers accompanied them every step of the way. At the Restoration of the Monarchy in 1660 the Pendrill brothers received gifts and annuities, while the pension bestowed upon them is still paid to their descendants in the twenty-first century.

The theme of horseracing inspired the naming of Kempton Drive from Kempton Park, Surrey; Lambourn Close, named after Lambourn, associated with the horses since the days of the famous White Horse at Uffington; Lingfield Drive and Lingfield Close remind us of the popular Surrey racecourse which sees the Derby Trial raced in early May each year; and Newbury Close is named after the Berkshire course which saw the first race in 1805 but which has only been held on this site for just over 100 years.

Naval vessels were the inspiration for another group of roads, in turn named after other famous names. Anson Close and Anson Road are named after HMS *Anson*, a Second World War Royal Navy battleship which was named after Admiral George Anson, 1st Baron Anson (1697-1762). He circumnavigated the globe and earned his reputation fighting the Seven Years' War.

Colinwood Close is missing a letter for it is named after HMS *Collingwood*, a Royal Navy stone frigate of 1940. The term 'stone frigate' is used to describe a training establishment which is not on ship and will never be afloat. Lord Collingwood (1748-1810) was a distinguished admiral at the turn of the nineteenth century. He fought alongside Horatio Nelson in several victories of the Napoleonic Wars.

Anson Close is named after the family who are buried in the crypt of Colwich Church.

Ramilles Crescent recalls HMS *Ramillies* and, once again, has not been spelled correctly. This revenge class battleship of the Royal Navy was named after the Battle of Ramillies, a major engagement of the Spanish War of Succession fought 23 May 1706.

Achilles Close, after HMS *Achilles*, a Leander class frigate of the Royal Navy launched in 1968 and which saw service in the Far East and Africa. Decommissioned in January 1991, it was sold to the Chilean Navy and, as far as we are aware, is still patrolling Chile's long Pacific coastline.

Ajax Close is after Leander-class light cruiser, HMS *Ajax* of the Royal Navy. During the Second World War, this vessel saw some of the fiercest action on the high seas with roles in the Battle of River Plate, the Battle of Crete, the Battle of Malta and the Battle of Tobruk.

HMS *Frobisher* was the inspiration for Frobisher Close, with the Royal Navy heavy cruiser patrolling the waters of the Indian Ocean and the African coastline, before playing a role on Sword Beach in the D-Day Landings.

Norfolk Grove is after the Royal Navy heavy cruiser HMS *Norfolk*. Travelling to the Americas, West Indies and East Indies, the vessel was undergoing a refit in the dockyard when the Second World War broke out. In 1941, HMS *Norfolk*, HMS *Rodney* and HMS *King George V* were the vessels which sank the famous German battleship *Bismarck*.

Orion Close is named after HMS *Orion*, which saw most of the Second World War in the Mediterranean, including the evacuation of Crete when she was badly damaged and 300 lives were lost.

HMS *Triton* gave a name to Triton Close. This vessel was a submarine of the Royal Navy, named after the son of Poseidon and Amphitrite in Greek mythology. Serving in the Atlantic Ocean and Mediterranean Sea, HMS *Triton* was sunk in the Mediterranean on 18 December 1940, although the circumstances behind the sinking are confused and uncertain.

Finally Leander Close, which takes the name of the character in Greek mythology who swam the Hellespont every night to be with his lover Hero but was drowned when bad weather beat him. Meanwhile HMS *Leander* joined the NATO task force patrolling the waters off apartheid-governed South Africa, later giving support to British fishing boats during the so-called Cod Wars with Iceland.

Themes in Great Wyrley include a general theme of plants. This is seen in Bluebell Lane, Fair Oaks Drive, Lilac Lane, Plants Close, and Poplar Road. Acorn Close is cut off and derived from Oak Avenue, and it is one of a number of names in Great Wyrley which have taken the popular woodland message. This cannot be called a theme as such for the roads are from different regions and are not all contemporary. Such can be seen in Ash Lane, Beech Court, Cedar Grove, Forest Way, Fern Drive, Honeysuckle Way, Lime Close, Oaken Grange and Pine Close.

Weaving is an original theme and gives Bobbin Close, Loom Close, The Spindles and Weavers Close. Beaumont Close is from Beaumont Road, one of a number of roads which are obviously derived from a personal name. However, many such street names are never recorded, and thus origins are unknown for Hartwell Lane, Martin Grove, Trevor Avenue, Wardles Lane and Penny Court.

As with other parishes and towns, there are always those names with meanings or origins which are either difficult to understand. They appear to have no tangible link to the place itself, and to simply been chosen arbitrarily. For example, Alpha Way features the first letter of the Greek alphabet and is a name often chosen as being representative of 'the first'. However, this does not seem to fit here unless there were once ideas of further development in the arca.

Alwyn Close surely cannot be named after conductor Kenneth Alwyn, conductor and composer William Alwyn CBE or former Archbishop of Wales Alwyn Rice Jones. The only other known origin is the Alwyn field in the North Sea which provided oil and gas from the 1970s. It remains to be seen if the coming of the then so called 'natural gas' to the area coincided with this development and provided the name.

Appledore Close may indeed have been named after the village in Devon near Barnstaple, or others in Devon, Cornwall or Kent, but just why such a name would be chosen is unknown. Cherrington is a village in Shropshire and the most likely inspiration for Cherrington Drive. Jubilee Close most likely refers to the Silver Jubilee of Elizabeth II in 1977. Whilst there is no evidence, Hut Hill Lane must be a reminder of a local name even if not an official name until later.

Saxon Close is a modern development with an ancient name, and Shanklyn Close takes its inspiration from the place on the Isle of Wight. Southfield Way seems to come from the landscape, yet as no record of any earlier name remains it may have simply been created to appear a former field name and fit in with the rural idea. Indeed, this may also have been the case with Spring Drive and Oxley Close. Streets Lane is a name which most often refers to an Old English street and a Roman road. However, this street, not from that era, is more likely to have been derived from a personal name, especially as it may have the remnant of a possessive 's'.

Peacock Croft may well be an example of a developer's name, for it has no historical significance. Strawberry Lane, too, was probably named at a developer's suggestion, coined to convey an impression of the finer things in life. Yems Croft is an odd name for it appears nowhere else. Having no traceable etymology means, Yems Croft must have been created solely for this place and, without any record of the reasoning behind it, cannot be explained.

# Chapter Seven

# Hammerwich—Huntington

## Hammerwich

Lions Den was, for most of its history, known as Elder Lane. However, by 1881 it had been renamed and can only have come from Thomas Lyon who lived here from the 1840s. It seems this was not named for any reason other than that Thomas Lyon's Den had a particular ring to it, for Lyon did nothing warranting a mention during his lifetime.

Ashmall is impossible to define with any certainty. Different spelling makes it difficult to know where this originates. In 1834 William Ashmale was a yeoman farmer in Edial and John Ashmale was a farmer in Woodhouses. By 1851 Joseph Ashmall (note the change of spelling) is seen as a farmer in Edial, and is the most likely origin of the street name. The sole record from Hammerwich is from 1662 and is of one Elias Ashmole.

Mansion Drive recalls the large house that Simon Biddulph of Lichfield built. The first version was built in the mid-sixteenth century. During the 1780s the house was rebuilt, and it was again a century later when it was known as Hammerwich House. From 1941 it was an outpost of Birmingham Women's Hospital and five years later a remand home for girls. By the 1960s it was a children's home and in 1985 a nursing home for the elderly. So many changes have led to this being known locally as Mansion Hill, and when the road was developed in 1988 it quite rightly took the local name.

Mill Lane leads to the old Speedwell Mill. The oldest datable windmill in the whole county, it has been here since 1300 when the miller was Henry Wymer. Yet while Henry is the earliest documented miller, it is also on record that he inherited this and three other mills from his father in 1280. However, none of the other windmills are datable and thus while this marks the site of the earliest known windmill in the country, we can deduce that it was by no means the first.

This would have been one of the first windmills in the land; wind power was state of the art technology in the late thirteenth century, prior to which mills were powered by moving water.

Church Road refers to St John's Church, and Station Road recalls the railway which came to Hammerwich in 1849, from 1959 was an unstaffed halt and closed in 1965.

Blackroot Close is named after Blackroot House which stood here, the abode of T.B. Wright, and itself named after a herb which has a pungent root used to make tea, can be brewed and may also be chewed and even eaten. Forge Close was the site of the old blacksmith's home.

Recently developed roads here have taken an avian theme. Refreshingly, these are a little more imaginative than the normal offerings. Fieldfare Close is taken from the member of the thrush family *Turdus pilaris*, which has an Anglo-Saxon name referring to 'the traveller through the fields', an accurate description of its constant foraging habits. Siskin Close is named from the European siskin *Carduelis spinus*, a member of the finch family which inhabits coniferous woodland. About six inches in length it is a seed-eater, but will take insects during the breeding season when it produces two or as many as six eggs. The siskin belongs to a group of birds known as passerines of the order *Passeriformes*, commonly referred to as either perching birds or songbirds. To say these birds are successful is an understatement, for more than half all the birds on the globe are passerines, one of the most diverse orders of vertebrates in the world.

Redwing Close comes from the bird of the thrush family *Turdus iliacus* native to Europe and Asia. The bird derives its name from the red underwing, is some eight inches in length, is migratory, breeds in conifer woodland where it nests in the underbrush or even on the ground and laying a clutch of four to six eggs. As with many thrushes the bird is omnivorous, taking berries, insects and earthworms depending upon the season and its migratory habits. There are over forty million individuals in Europe alone and is not considered endangered, although numbers can crash dramatically in the severest of winters and wettest of summers, but will recover quite quickly.

Hammerwich Link is a comparatively modern road cut to accommodate the train line and to link the two halves of Hall Lane. Hall Lane is named after Hammerwich Hall, which was the home of William Heath in the eighteenth century. Burntwood Road is a destination name, while Middleton Close and Overton Lane refer to two minor place names first found in the fourteenth century.

Coppy Nook Lane is not known to have been inhabited before the late eighteenth century. Many different ideas have been proposed for the first element here: *copp* 'summit', the personal name Coppa and 'copper', referring to the colour rather than the metal itself. However, the author feels the clue lies to the east of Hammerwich and Coppice Lane. It is easy to see how one lane could have existed here passing directly through Hammerwich and probably existing well before it. A change of name for that part of the lane where the settlement grew was to be expected, thereafter either of the two halves were open to corruption. Thus it is quite likely this is 'the nook of land of or by the coppice'. Pingle Lane is from the Middle English *pingel* or 'a small piece of enclosed land'. Meerash Lane an old field name telling of 'the boundary of ash trees', and Green Lane is blanketed by vegetation.

Hanney Hay Road this is undoubtedly a minor place name or a field name, but the place is unknown and assumed to have been renamed or lost. It does not help that the meaning of 'Hanney' is obscure, although 'Hay' is an administrative division of a forest. With Stockhay Lane normally the 'Hay' part cannot be an administrative region of a forest – it is far too small. In combination with the first element this could well be 'the place where hay was stored', clearly a barn, but as it is not named as such it suggests this was a communal storage facility and probably for bedding rather than feed.

## Harlaston

Main Road is found is most towns and villages and speaks for itself. Manor Lane, and off that Manor Close, remind us of the lane to Manor Farm and of the old manor itself. The modern but tasteful development of Churchside is alongside the modern church but stands on the site of the former wooden church.

Haselour Road is, strictly speaking, a destination name, although it effectively leads past the splendid Haselour Hall. This hall was first built by the Timmon, but the Arderne family stamped their name on the place. Sir Thomas Arderne died in 1391 after leading the men of Haselour and neighbouring Elford in victories against the French at Crecy and Poitiers. In the fifteenth century the Arderne heiress married St Thomas Stanley, a man reputed to have given the Earl of Richmond (soon to be Henry VII) a bed for the night on his march from Lichfield to the Battle of Bosworth.

## Haunton

By the time Main Road of Harlaston has reached Haunton to the north-west it has become Main Street, but is still the principal road in the village. Mease Lane is named from the River Mease, still the dominant feature in the landscape here. The river name means 'the mossy one' and is, historically, an important river for it forms part of the borders between Derbyshire with Leicestershire and of Leicestershire with Staffordshire. Rising in the Leicestershireshire coalfields it flows for sixteen miles before joining the Trent with the Tame near Croxall Lakes. Syerscote Lane is the sole destination name in the village, although this is a gated road and Syerscote itself is little more than the collection of buildings it always has been. Finally Twizles Lane must be a reminder of a former resident, yet there is no record and it could well be so badly corrupted as to be completely misleading. Furthermore, Twizles Lane no longer leads anywhere.

## Hazel Slade

Bentley Brook Lane takes its name from the nearby Bentley Brook. New Hayes Road is adjacent to 'the new woodland' from which it takes its name. However, it is unclear if the names of Barnfield Way, Briars Way, Deerhurst Rise, Old Park Road, Slade View Rise, Summerside Avenue, Valley Way and Warren Close recall names which already existed before development or if they were created in order to appear to be typical 'countryside' names.

Sevens Road ran from the Number Seven Pit of the Cannock Chase Colliery Company and was known locally as simply Sevens. It can probably be considered a destination name, yet not in the normal way. This is also the case with Littleworth Road, Rugeley Road, Cannock Wood Road, Rawnsley Road and Cannock Wood Street. A connection, rather than a theme, links the streets of Eastgate, Westgate and Middleway, as each describes the relationship to one another rather than strict compass points.

Beaudesert View tells us it overlooks the estate of Beaudesert, home of the Paget family who also proved the inspiration for Pagets Chase. Longstaff Avenue is named

The River Mease has provided the name for many streets, here flowing out of Harlaston.

after the family of Mrs Grace Longstaff, a hosier in Station Road, Hednesford, at the end of the nineteenth century. Other streets named after individuals in Hazel Slade, be they associated with the coal mining industry, merchants, landowners or local characters, include Chetwynd Park, Danby Drive, Glovers Close, Jesmond Close, Saunders Close and Williamson Avenue.

A theme of saints has given us St Andrew Close, named after the patron saint of Scotland, Russia, Sicily, Greece, Romania, mariners, fishermen, fishmongers, rope-makers, singers and performers; St Bernards Close, the patron saint of mountaineers, skiers and the Alps; St Christopher Close, patron saint of bachelors, travellers, storms, epilepsy, gardeners, and toothache; St David Close, the patron saint of Wales, vegetarians and poets; St George Close, the patron saint of England, Netherlands, Germany, Lebanon, Bulgaria, Catalonia, Constantinople, Ethiopia, Georgia, Malta, Palestine, Portugal, shepherds, soldiers, lepers, horses, farmers, equestrians, archers, butchers, skin diseases, herpes, and the Brazilian football club Corinthians; St Francis Drive, the patron saint of animals, merchants, San Francisco, and stowaways; St Matthew Close, the patron saint of accountants, Salerno and Italy; St Michaels Drive, the patron saint of paratroopers, police officers, mariners, grocers, and paramedics; St Patrick Close, the patron saint of Ireland, Nigeria, New York, engineers, and Boston; and St Thomas Drive, the patron saint of architects and India. Sadly the theme could not decide whether to include the possessive 's' or not and both are present.

Aintree Close, Goodwood Close and Kempton Close all reflect the theme of horseracing. One more is Jenkinstown Road. This horse was described as having to be trained as hard as four horses and was never fancied in any of his races. Somehow the horse found itself at the tape at Aintree and the world's most famous steeplechase and, four and a quarter miles later, was in the winners' enclosure, victorious. Jenkinstown entered the race twice more and never finished, indeed the horse never tasted victory again.

Trees are undoubtedly the most popular theme. Here we find Alder Way, Aspen Court, Cypress Rise, Hazel Drive, Hornbeam Crescent and Juniper Close.

## Heath Hayes

Local place names have always been the principal source of the earliest roads. Newlands Court and Newlands Lane remind us of Newlands Well, here in the middle of the nineteenth century. Bank Street may not be so clearly defined as a ridge of ground today but would have stood out when the region was less developed and is taken from a note on an old map. Similarly it seems the name of Boston Close has taken the name shown on an old map and not, as we might think, the well known US city or the Lincolnshire town. Cross Street allows one to cross between Bank and Stafford Street.

Destination names are the natural source for the longer roads. Here we find Cannock Road, Norton Road, Hayes Way, Hednesford Road and Wimblebury Road.

Individuals who have contributed to the history of an area are undoubtedly deserving, perhaps more so even than famous individuals, to be honoured by a street name. Indeed, there would doubtless be more streets named after influential locals, were it not so potentially controversial. Mr C. J. Hodson, an agent for the Paget estate, gave his name

A path through the beauty that is Cannock Chase.

to Hodson Way. Gladstone Road can only refer to the former prime minister William Gladstone (1809-98). A member of the Liberal Party, the use of his name here seems purely arbitrary.

Heligan Place is an unusual and imaginative name which can only have come from one source. The Heligan Estate was the owned by the Tremayne family since the sixteenth century until it was sold shortly after the First World War. Located near Mevagissey in Cornwall the impressive botanical gardens were long forgotten and neglected, largely through sixteen of the twenty-two gardeners becoming fatalities of the Great War. In 1996 the gardens were restored as part of a Channel 4 series, which also resulted in the gardens providing local employment once again. Clearly the project made quite an impression on someone in the Heath Hayes area at the time.

Themes are popular in the area for there are a number of more recent developments. Trees are represented by Acorn Close, Almond Close, Apple Walk, Chestnut Close, Elder Close and Hickory Court. Precious stones are seen in Amber Grove, Diamond Grove, Jade Grove, Sapphire Drive and Turquoise Grove. British prime ministers are honoured by Asquith Drive, Attlee Grove, Baldwin Grove, Callaghan Grove, Lloyd George Grove and Wilson Grove. Composers were chosen for Chopin Close, Greig Court, Handel

Court, Mozart Court, Purcell Close, Strauss Drive and Vivaldi Drive. A second dip the theme of the arts brought streets named after famous painters: Goya Close, Picasso Close, Rembrandt Close, Turner Close, and Van Gogh Close. The Brontë sisters were the inspiration for Bronte Drive; however, there are no other writers here and this probably should generally be considered a region of famous names.

Staffordshire place names are covered by Blithfield Place, Ingestre Close, Shugborough Way, Trentham Close and West Close. Spode Place can hardly be considered a place name, but it does remind us of the famous Staffordshire pottery. English counties are always a favourite choice and here is no exception with Dorset Road, Kent Place, Rutland Road and Stafford Street. London place names include Buckingham Place, Grosvenor Road and Kensington Place.

The net was cast across as far as Australia to find the names of Darwin Close, the original city named after Charles Darwin; Hobart Road, the original Hobart, the capital of Tasmania, was named after English Secretary of State Lord Hobart who founded the city in 1804; Melbourne Crescent and Melbourne Road, were renamed in 1837 in honour of English prime minister Lord Melbourne, prior to that it was known as Dutigala and Heath Hayes residents will doubtless be pleased the name was changed. Canada, another Commonwealth country, provided the inspiration for the theme which includes Edmonton Close, the original city was named by William Tomison of the Hudson Bay Company whose clerk, John Peter Pruden, was born at Edmonton near London. Hudson Close takes its name from Hudson Bay, in turn named after Henry Hudson (1550-1611) who discovered the region in 1610; Huron Close, named after Lake Huron, itself named after the Native American Huron Tribe who were called such by the French settlers on account of their bristly hair, said to be like that of a boar (French *hure* or 'boar's head'); Michigan Close (referring to the lake, not the US state) is derived from the Native American Chippewa word *michaw-sasigan* and meaning 'great lake'; and St Lawrence Drive, the name of a river, gulf and man-made seaway first explored by French navigator Jacques Cartier in 1534, who named it such for it was on the feast day of St Lawrence (10 August) when he first sailed into this area.

For the theme in another area look to the Lake District and the names of Rothbury Green, a place name meaning 'Hrotha's stronghold'; Kielder Close, from a Celtic river name meaning 'the rapid stream'; Alston Close, a place name meaning 'Halfdan's farmstead'; Eden Close, a Celtic river name meaning simply 'water'; and finally Alnwick Close, a place name meaning 'the dwelling on the River Aln', which is an uncertain Celtic river name.

Nearby Langholm is taken from the Scottish town of that name, which is also known by the strange name of Muckle Toon and residents must be delighted this was not the chosen street name. Closer to home and the delightful county of Devon gave names to Ansty Drive, Langtree Close, Buckland Close, Otterburn Close, Brampton Drive and Denbury Close. Then off to Shropshire for Attingham Drive and Hodnet Place. Lyndhurst Road features a place name in Hampshire's New Forest, but could also be a personal name. It is often difficult to tell if street names reflect place names or surnames. Such is the case with Cleeton Street, possibly a place in Nottinghamshire; Ludlow Close, Tutbury Close and Hartlebury Close, which could feature those Worcestershire towns; while Newhall Close and Wootton Close feature place names which are found in several counties. Statistically the majority are likely to be after place names, yet without any known candidates it is impossible to suggest which (if any) are taken from surnames.

A theme of cathedral cities has created Bristol Close, Canterbury Way, Chester Close, Chichester Drive, Ely Close, Gloucester Way, Peterborough Drive, Rochester Way, Salisbury Drive and Truro Place. While St Paul's Close refers to the best-known cathedral in London by name.

Bird names have been transferred to street signs for Capercaillie Drive, Cuckoo Close, Firecrest Close, Grouse Way, Hobby Way, Nightjar Way, Nuthatch Close, Partridge Close, Peregrine Way, Pheasant Way, Quail Close, Robins Croft, Sparrowhawk Way, Whintchat Drive, Woodpecker Way, Woodpigeon Drive and Wrens Croft. However, the following birds are a separate theme, for these names are all those of birds of prey or raptors: Condor Grove, Eagle Grove, Hawks Green Lane, Kestrel Grove and Osprey Grove. Plants and specifically the wild varieties have been used for Charlock Grove, Marigold Close, Primrose Meadow, Rosebay Meadow and Willowherb Close.

Animals put in a brief appearance with Badgers Way, Squirrel Close, and even Foxhill Close despite the addition. However, Badgers Court, some distance from the three aforementioned streets, is the name given by a modern developer. Similarly Clover Meadows, Birchfields Drive, Gorsemoor Road, Avenue Road, Burdock Close, The Coppice, Edgemoor Meadow, Fairfiield Close, Green Meadows, Harebell Close, Heathland Close, Heath View, Heath Way, Hemlock Way, Hopton Meadows, Hyssop Close, Lawnswood Close, Meadow Way, Redbrook Close, Shire Hall Place, Spindlewood Close, Sweetbriar Way, Thistledown Drive, Wheatlands Close and Woodford Way were all names given by modern developers.

The Heath is hardly so much as wild land today but the well-manicured grass of a sports field. Highfield Road may be either an old field name or a street name created to appear as such; however, there can be no doubt as to the incline of Hill Street. Sidon Hill Way seems to be named after a minor place name, and St Johns Court has connections to the dedication of that saint.

There are always names that, for one reason or another, are not recorded and as a result the meaning and origins are either unknown or misunderstood. Carlton Close and Barber Close (and Sam Barber Court) are almost certainly named after individuals, but just who is unknown. Bourne Close could originate from a family name, but it is more likely to have been named by a modern developer and created hoping to suggest a desirable countryside location to prospective buyers. Brooklyn Road is another name with two possible origins: either a personal name or the well-known part of New York City. Chapel Street was the site of a chapel, but Chaplain Road seems more likely to refer to a person than the religious post. Charterfield Drive does not seem to have any etymology and may be a created neutral name. Deavall must be a personal name, although no record of this surname is found around this part of Staffordshire.

Coal Haulage Road has obvious connections with the mining industry, yet also seems to be a traditional name. Cromwell Road probably does not refer to the former Lord Protector of England, but, more likely, it refers to one of that family who were major landholders in Staffordshire following the Restoration of the Monarchy. Holston Close is a street named with mysterious etymology. Holt Crescent, however, is undoubtedly an old landscape name. Mill Crescent and Millers Vale seem to have been created to give a rural image to a suburban street. Nicholls Way and Tranter Crescent are undoubtedly named after individuals, although their identities are unclear.

## Hednesford

As always local names have proven an inspiration for many of the oldest street names. Belt Road refers to the Huntington Belt, a strip of land on 'the huntsman's hill'. Bradbury Lane, and from that Bradbury Court, marks the position of an earthwork. Brindley Heath gave its name to Brindley Heath Road, Brindley Road and Brindley Crescent. Littleworth Hill and Littleworth Road derive their names from the place name meaning either 'the poor quality enclosure' or 'the small enclosure'.

There were once quarries in this area and, while there is no official record of any lime being extracted, Limepit Lane is a good clue that it was associated with the oxide of calcium used in the smelting of iron ore. Market Street, previously Station Street, was named such in the first decade of the twentieth century. Farm Close predictably points to the farm here.

Broadhurst Close comes from Broadhurst Green Road, itself from a minor place name, while a well-known sight gave us Cottage Close and Cottage Lane. Mount Street, Mountside Street, Mount Avenue and High Mount Avenue all refer to these elevated sites. It makes sense to name a road after its destination – it is a natural signpost. Here we find Cannock Road, East Cannock Road, Old Hednesford Road and West Cannock Way.

The Paget family's influence has named several streets, the title of Marquises of Anglesey being the origin of Anglesey Street, Anglesey Street and Anglesey Crescent. The family also held the title Earls of Uxbridge, hence Uxbridge Street and Uxbridge Court. John Ward was granted an alehouse licence in 1604 where Ward Street is today. Councillor Stanley, who held office during the expansion era of the 1890s and was also agent for the Chase Coalfield, deservedly had Stanley Road named after him. The Hearth Tax Records of 1666 show Edward Stringer, Henry Rowley and Thomas Beech, who provided the inspiration for Stringers Hill, Rowley Close and Beech Court respectively. Levetts Hollow is named after the family who were major landholders in this area. Cotton Grove is named after owners of the Blue Yard Brickworks in the 1840s, where Cotton Grove is cut today.

Forge Street is found in an area best known for the training of horses, and it was here, at Hill Top, that John and George Benton ran a forge in around 1845. Florence Road reminds us of Florence O'Driscol, the one female horse trainer, and Eskrett Street, trainer Thomas Eskrett. Farmers Samson and James Blewitt gave the name of Blewitt Street. While Bakers Way may seem to point to a man of that trade, it is actually named after Joseph Baker, a local businessman. Reservoir Road recalls the large volume of water which was being created when, at 2pm on Sunday 11 June 1887, one bank collapsed and sent water cascading though the gap. Damage was minimal but the project never saw the light of day again, although it was not officially abandoned until 1930.

Fieldhouse Road has insufficient records to show if this is from 'the house in the field' or if it is – equally plausible, albeit less likely – 'the house of the Field family'. Abbey Street is on land owned by the Reverend Richard in the mid-nineteenth century. Walkers Rise comes from the nineteenth-century field name of Walkers Piece.

Broadway certainly lives up to its name. Shaftesbury Drive remembers the Earls of Shaftesbury, the Howard family, who hold the title of Dukes of Norfolk, the highest dukedom in the land. The Howards, one of the most famous families in England's history, also gave their name to Howard Crescent.

Themes in Hednesford include place names of Scotland. These are seen in Ardgay Drive, stemming from a Gaelic name meaning 'the height of the wind'; Kenmore Avenue, from the same language and meaning 'big head', a topographical description; Kinross

Avenue, from the Scottish Gaelic for 'the head of the promontory'; Arnotdale Drive, from the '(place of) the earth nut' and referring to pig nuts; Balmoral Drive, from 'the homestead in the big clearing'; Braemar Gardens, from 'the upper region of the place of Mar' – it remains to be seen if this is an old place name or a personal name; Melrose Drive, from the '(place at) the bare moor'; and Linwood Drive, from 'the wood by the pool' and a hybrid of Brythonic *Llyn* and Old English *wudu*.

A number of names were coined to identify with the overgrown countryside present before the developers moved in. These include Briar's Close, Evergreen Heights, Heather Mews and Heather Road. Birds are always a popular theme, here seen in Chaffinch Close, Meadow Lark Close, Herondale and Swallowfields Drive. However, when looking for a national hero, nobody has more pubs named after him than Lord Horatio Nelson, and thus he and his career make a good subject for a series of names: Nelson Drive, Trafalgar Close, Victory Close and Hamilton Close. Mirroring the battle of Trafalgar is that of Waterloo Boulevard.

Much has been written about Nelson's achievements and quite rightly so, for one would go a long way to find an Englishman who has had more influence from a military perspective. Nelson's titles are inscribed on his coffin as 'The Most Noble Lord Horatio Nelson, Viscount and Baron Nelson, of the Nile and of Burnham Thorpe in the County of Norfolk, Baron Nelson of the Nile and of Hilborough in the said county, Knight of the Most Honourable Order of the Bath, Vice Admiral of the White Squadron of the Fleet, Commander in Chief of His Majesty's Ships and Vessels in the Mediterranean, Duke of Bronte in Sicily, Knight Grand Cross of the Sicilian Order of St Ferdinand and of Merit, Member of the Ottoman Order of the Crescent, Knight Grand Commander of the Order of St Joachim.' The King of Naples created him Duke of Bronte in July 1799, leading to him signing his name as 'Nelson & Bronte'. Also he was a Colonel of the Royal Marines, the University of Oxford bestowed upon him the honorary degree of Doctor of Civil Law, while he was granted the Freedom of the City by Bath, Exeter, Hereford, Monmouth, Oxford, Plymouth, Salisbury and Worcester. Sadly he had no legitimate heirs and thus most of his titles died with him in 1805, although there were moves made to ensure that his brother, the Reverend William Nelson, was created Earl Nelson, Viscount Merton of Trafalgar and Duke of Bronte.

One of the most famous landmarks in the world is Nelson's Column. It was scaled for only the third time in its 160-odd year history in 2006, for cleaning and general maintenance. Using modern measuring techniques it was discovered that the 185 ft monument was in fact only 169 ft high.

Fisher Street marks the home of Edwin Fisher, Hewston Croft is in honour of draper William Hewston, bookmaker Joseph Marston is remembered in Marston Road, racehorse trainer Thomas Metcalfe produced Metcalfe Close, Stevens Drive comes from beer retailer Charles Stevens and Moore Street could refer to outfitter William Moore or licensee Jane Moore of the Lamb and Flag. Denmark Close was cut on land which was previously associated with Denmark Villa. Keys Close and Keys Park Road mark the site of the Cross Keys, a historic coaching inn.

Victoria Street and Albert Street mark the boom period of the reign of Queen Victoria with her consort Prince Albert; other royal houses are honoured by the street names of Tudor Road, Windsor Court and Windsor Avenue. Booth Street remembers William Booth, founder of the Salvation Army (there is a branch near here); Ebeneezer Street, Wesley Place and Kingsley Street are associated with the Methodist Church. Poets Rupert Brooke and Thomas Gray gave their names to Brooke Road and Gray Road.

Richard Cobden's work as a statesman and champion of the poor is well documented and definitely merits the naming of Cobden Road after him.

Councillors have been honoured by the naming of the following: Clarke's Avenue, Foster Avenue, (Don) Metcalfe Close, Millicent (Rowley) Close, Moore Street, Simcox Street, (Jack) Sunley Drive, Wilcox Avenue and finally Walkers Rise, after Nancy Walker who was once a common sight on the Chase when walking her goats. Lomax Road marks the work of local authority engineer Eddy Lomax.

Pye Green Road, Rawnsley Road, Rugeley Road and West Cannock Way are destinations. Stafford Lane, which probably began its life as a destination name, has been styled an abbreviated form of the name of the county and added to by the names of Bedford Place, Essex Drive, Hereford Road, Cornwall Road, Radnor Rise and Sussex Drive. The Chase is a favourite subject, even when it is referred to indirectly, and Chaseside Drive leads to the Chaseside Industrial Estate. Lower Road does lie at the lower part of the hill, and Millpool Road is from the old millpond here in the mid-nineteenth century. Passfield Avenue remembers a nineteenth-century farming family, and New Street has not been 'new' for many years. Western Road and West Hill Avenue are on West Hill.

The unusual theme of baking seems to have been behind the names of Bakers Way, Bailey Close and Granary Way. Communication is the key to Bell Drive, Marconi Place and Edison Close. Church Hill marks the road to the parish church of St Peter's and St Peters Road. Kings Avenue leads to Kingsmead High School, from which it takes its name. Priory Road has never been held by any religious body and is likely therefore named to mark nearby Church Hill. Railway View does indeed stand close to the station, as do The Sidings and Station Road, while Stephenson Way honours the builder of the famous *Rocket* locomotive. Splash Lane would have been wet for much of the time, the problem accentuated by constant use causing ruts and hollows to develop and allowing water to accumulate. Tower Road is dominated by views of the tower which gave its name.

Place names often provide a well of ideas for street names, although why these places are chosen is never likely to be known. Included among these names are Charnwood Close, 'the wood in rocky country'; Croxley Drive, 'Krokr's woodland clearing'; and Sherbourne Avenue, '(place at) the bright or clear stream'. A series of 'hilly' names are seen in Cotswold Close and Cotswold Road, Cherry Bank, Cheviot Rise, Glendene Road, Hillside Close, Mendip Road, Pendle Hill, Rosehill, Springfield Rise, and, just to offer the opposite point of view, Valley Road.

Modern developments often have names which have no etymological basis. These are simply created to raise the profile and make it convey a more pleasant idea than a name, such as, for example, Slum Alley would. Examples include Alpine Drive, Berry Hill, Beverley Hill, Bluebell Close, Bond Way, Bracken Close, Bramble Drive, Cedar Close, Common View, Buckthorn Close, Cygnet Close, Daisy Bank, The Dell, Foxglove Walk, Foxtail Way, Glendene Road, Gorse Way, Green Heath Road, Greensalde Grove, Greenwood Park, Heather Valley, Hedgerow Close, Leafdown Close, Linden View, Longacres, Maycroft Close, Meadowsweet Way, Mountain Pine Close, Pine Tree Close, Plantation Road, Sandpiper Close, Sunrise Hill, Thornhill Road, Conifer Close, Beech Pine Close, Corsican Drive, Silver Fir Close, Spruce Road, Stone Pine Close, Viewfield Avenue, Waterlily Close, View Street, White Bark Close, Whitethorn Close and Woodland Close.

Some residential areas take developer's names, have no etymological basis and are named, partly, to provide a point of reference during planning and construction stages. It is difficult to imagine such names being created for any specific reason, other than

The rookery which gave Rookery Road its name is hidden by the leaves, but the birds themselves are very audible at Hints.

being a good idea at the time (this is not a criticism, merely an observation). This is the case with Bryans Way, Pope Grove, Burleigh Close, Cardigan Place, Carmel Close, Chalcot Drive, Chancery Drive, Charlemonte Close, Bromley Close, Cowley Green, Croft Avenue, Prospect Manor Court, Raven Close, Rydal Close, Sheraton Close, Stagborough Way, Wyvern Grove, Copperkins Road, Coppermill Close, Fisher Street, Heath Street, Hewston Croft, Meadway Close, Midhurst Drive, Mitcham Close, Sharon Way and Telford Grove.

Of course, there are always those names which have defied all attempts to discover their origins. Here McGhie Street, Marina Crescent, Marston Road, Mary Street, Mavis Road, Stevens Street, Telford Grove and Whitfield Road all come under that heading.

## Hilton

Cranebrook Lane is taken from the river name telling us of 'the brook where cranes or heron are seen'. Bullmoor Lane is 'the moorland where bulls are pastured'. Thorneyhurst Lane, while clearly a place name, leaves no surviving record of its origin, but it must stem from 'the thorny hill'. Clearly, from defining just three place names – albeit using a little imagination – it is possible to draw a fairly accurate picture of a place.

## Hints

Hints Lane and Hints Court take their names from the place name. Rock Hill is that part of the old A5, or Watling Street, which passes the point where sand and gravel pits were dug. Bangley Lane is an old landscape name also applied to the farm.

Rookery Lane leads to the hill of that name and is now covered in trees – making it even more prominent. The lane crosses a ford across the Black Brook. The river is channelled through a series of brick culverts, a reminder of the former splendour of the grounds of Hints Hall.

The ford at Rookery Road, Hints.

## Hopwas

Lichfield Road speaks of its destination, as do Hints Lane and Hints Road. Hopwas Hill has a name describing that it overlooks 'the alluvial enclosure'. The Old English term *wasc* or 'alluvial land' describes a region alongside a river which is regularly and seasonally flooded. In modern times this might not be particularly convenient but historically this was a veritable boon for the flood brought with it silt washed down from higher regions. This flood silt, or alluvium, is deposited on the land when the river bursts its banks and replenishes the mineral content.

The name of Two Trees Close comes from Two Tree Farm on which it was developed. The street name has the error and should not be plural. Plantation Lane originally led to and ran alongside the plantation but has become greatly extended over the years. Similarly Nursery Lane takes its name from a field. Dunstall Farm gave its name to Dunstall Lane, which, in turn, came from a now lost place name.

## Huddlesford

Thomas Capper gave his name to Capper's Field, and from there it was transferred to Cappers Lane. Brookhay Lane is from an old field name, Park Lane passed the grounds of the manor house, while Broad Lane may have been considered such at one time but is a distinct misnomer in the modern era.

Today, Huddlesford forms the north-eastern limit of the Wyrley and Essington Canal; work is currently ongoing to link it to the Coventry Canal.

## Huntington

Gratley Croft remembers Greatley's Piece, Staley Croft was Staley's Rough and had been a field name here for many years. Long Croft is an old field name. Common Walk was indeed common land two centuries ago, and Woodland Court is a reference to nearby Cavan's Wood to the east. Teddesley Way is an old and otherwise forgotten place name. Dual Way is aptly named for it parallels the A34. Pillaton Drive and Teddesley Way are named after former local councillors.

Limepit Lane was, in 1841, Stone Pit Lane and the site of a mine and/or quarry. Gravel Lane is shown as a gravel pit in the nineteenth century. Colliers Way is a reminder of the mining history of the area. Littleton Drive leads to the Littleton Business Park and to further mining memories and the colliery, which was at one time the largest in the Midlands. However, its beginnings were not the most auspicious for the original shafts, sunk in 1877, suffered from flooding to such an extent that new shafts were required within just twenty-five years. The mine employed 1,900 workers after extensive modernisation in 1982, only to close in 1994. Both the pit and the resulting spoil heap have been redeveloped for walkers and ramblers, and the area now fits in perfectly with the wilds of the Chase.

Cocksparrow Lane is an old name which may refer to sparrow clapping, for in some places sparrows were considered a source of food. Clearly many of the birds needed to be caught and so trappers would wait until the birds roosted for the night. Draping a net over the site of the roost and then scaring the birds by clapping, or, more likely, banging pieces of wood together, the sparrows would become trapped in the net and were easily removed.

Themes for more recent street names are commonplace. Tree names in Huntington include Almond Road, Ash View, Beech Grove, Birch Road, Cherry Tree Road, Elm Grove, Fir Close, Hawthorne Road, Lime Road, Maple Drive, Oak Avenue, Silver Birch Avenue, Sycamore Way, and Willow Walk. Birds provide another theme and here we find Linnet Close, Nightingale Close, Partridge Close, Raven Close, Redwing Drive, Robin Close, Skylark Close, Swallow Close and Woodstock Drive.

Wild plants of the chase is one of the better street ideas implemented in the area, and this theme has produced Bilberry Crescent, Bracken Road, Buttercup Close, Comfrey Close, Cowslip Close, Fern Road, Gorse Drive, Heather Drive, Lichen Close, Ling Road and Moss Crescent. A perfect compliment to the plant theme, streets named after creatures local to the chase are also to be found here. These include: Badger Close, Deer Close, Fawn Close, Squirrel Close and Stag Drive.

Other names that reflect the area are Chase Walk, of obvious derivation; Huntington Belt, a belt of land first named on a map of 1834; Stafford Road, a destination name. Wrekin View is one of the new developments alluding to the chase, as well as Huntsmans Rise, Foxfields Way, Holly Lane, Oaklands Close and Pear Tree Close. Meadow Croft is not an actual field name but a reminder of what was here before. Similarly, Ashdale Close is a suggestion of a more rural location. Three complimentary names here are Highland Road, Lowland Road and Midland Road.

# Chapter Eight

# Lichfield—Lynn

## Lichfield

Local names are the traditional sources of road and street names. Backcester Lane was, until the creation of the shopping centre, a part of Wade Street. The modern version is an odd corruption of Bakers Lane and its former name of Back Lane; indeed, it could be suggested that this name is unusual that it must be the result of a deliberate amalgamation of the two. Later the theme of the trade returned in the form of Bakers Lane. Ford Lane has been an important local crossing point for many years, with further evidence provided by Brook Cottages, also located here. Hobstone Hill Lane, rising to 149 metres above sea level, as a street name is recorded since the fourteenth century. Lying on the parish boundary, on or near a marker stone, the first element may be a hob (a sprite) or simply an old pet form of Robert.

Beacon Street was known as Bacon Street until 1836, indeed it had been for many centuries. It almost certainly came from that particular cured meat being produced or sold here. The change appears to have been simply a question of the desirability of the street name Beacon rather than Bacon, much as developers strive for ever more attractive names in the twenty-first century. From this name Beaconfields, Beacon Gardens and Beacon Mews were derived; Neather Beacon is lower and refers to itself as being such.

Beech Gardens was developed where a house called Beechcroft once stood. It was also chosen to mirror Elm Gardens, named for the number of elm trees around here, and it is probably also true to believe the house of Beechcroft was named for the beech trees which had stood there. Bell Close and Tuke Place are on the site of the former Tuke and Bell Engineering Company. Bird Street, which appeared as Byrd Street in 1506, is most likely named after a family who lived here. Blue Bell House is a recent project named after the Blue Bell Inn.

Boathouse Field takes its name from a field, which, in turn, was the site of the boathouse used by fishermen for access to Trunkfield Mill. Trunkfield Meadow was built on the site of Sandford Mill, and Alesmore Meadow is recorded as Alesmore in around 1800. Boley Lane is also a field name, which has spawned the newer developments of Boley Cottage Lane, Boley Close and Boley Park. Bore Street is first seen in 1506 and speaks of the boards laid on the bare and often muddy ground to support the goods being sold in the market there. Borrowcop Lane is accepted to have come from an Old English place name in *burgh-cop* for 'the fortified high place' although no record has even been found. Breadmarket Street has always been associated with the sale of bread.

Castle Dyke is believed to be the site of an early fortification in the city. Chapel Lane reminds us that the Catholic Church of St Peter ad St Paul was only considered a chapel when it first opened its doors to worshippers. The Charters reminds us of the seven charters granted to the City of Lichfield. The Chequers is named after a public house which was formerly here.

Bower Close is a well-chosen name for a Lichfield street, for it commemorates the annual festival of the Lichfield Bower. Held every spring bank holiday week, it dates back to at least the twelfth century when Henry II ordered that all men able to bear arms should report for inspection by the local magistrate, for, with no army, it was the only way to ascertain the nation's potential fighting force. A carrot of roast beef and wine was offered at a nominated bower house, so ensuring a good response.

Locally the magistrate would be under the control of the Court of Arraye and the bower house was erected at Greenhill and decorated with laurel and lilac. All men of sound mind and body between sixteen and sixty would parade through the city as if they were marching to war. By the seventeenth century, the Lichfield Bower (also called the Greenhill Bower, for obvious reasons) had been embellished by maids and morris dancers, flowers and flags, song and dance. As the mood off the nation changed, so did the Lichfield Bower; it adapted to reflect the gaiety of the monarchy or the more sombre attitudes of the commonwealth. When musketeers joined the procession in later years, the firing of a volley of shots over the houses of the leading dignitaries soon became a tradition. This signalled to the occupant that the procession had arrived and that cakes and ale were then to be distributed before they could pass on.

Thus the complexity and level of celebration changed with the passing centuries until, in the modern era, a whole weekend is dedicated to the Lichfield Bower. Local groups and businesses contribute in the form of stalls, games, dance and song, with the main focus of the jollity on the fairground. Still the procession passes through the town and awaits the permission of the Town Clerk at the Guildhall to disperse and enjoy the merriment. The alternative name for the Lichfield Bower is the Greenhill Bower, named from the place where the Bower House was erected and where the road is today named Greenhill. Greenhill is one of the oldest parts of the city and simply describes the appearance of the place when it was first settled. In recent times it has also been the basis for Greenhill Mews. It was also where the cattle market was held for many years, which gives the name of Smithfield Rise.

Broad Lane was very much wider than other roads of the day in the eighteenth century. Capper's Lane ran alongside Capper's Field, land owned by the Levett family and farmed by Thomas Capper in the nineteenth century. Levetts Fields is a name taken directly from the open land here; it recalls a family who lived in St Johns House. Meadowbank Road is not an actual field name but a description of the area during the nineteenth century. Cherry Orchard reminds us that in the eighteenth-century market gardening and fruit production contributed to the city's wealth. The dedication of the church produced the names of Christchurch Lane and Christchurch Gardens. St Michael's Church is alongside Church Street.

Claypit Lane is a remnant of the ancient road network and was where clay was dug for making bricks. Conduit Street had a ditch dug to drain water from the area. Covey Close shows us that birds, probably partridges, were shot here in the early nineteenth century, but Crane Fields more likely refers to where heron were often seen. The Crossings marks where a level crossing was found. Cross Keys is predictably the site of a former public house, while Cross Lane literally crossed the road to Tamworth and Whittington and was a part of the Roman Ryknield Street. Thacker Drive occupies the area where Thacker's Barn stood in the eighteenth century, the property of Thomas Thacker two centuries earlier. Blakeman Way is a region said to have been held by Richard Blakeman the elder and Richard Blakeman the younger, while a John Blakeman is also mentioned in the tithe awards of the nineteenth century. The same records show Edward Maddox and Thomas Maddox being in residence near what is now Maddocke Walk. James and Samuel Whitehouse gave their family name to Whitehouse Drive, while Parnall Avenue takes its name from Parnell's Field on the nineteenth-century map.

Dam Street reminds us of where the water flow to the mill was controlled. Dark Lane is a common name for a particularly overgrown route. The Dimbles, Dimbles Hill and Dimbles Lane are all derived from a term for a dried ditch, once carrying water and later used as a footpath. Dovehouse Fields marks where a long-standing dovehouse was seen, which probably indicates possession by the church. Furnivall Crescent is named after Francis Dorothy Furnivall, who built a school in Stowe Street in 1833 and paid for a schoolmistress to teach the children of the poor.

Essington Close and Wyrley Close are on an industrial estate built on the line of the Wyrley and Essington Canal. In 1792 an Act of Parliament permitted the building of a canal linking Walsall and Wolverhampton to the coalfields of Wyrley Bank and Essington via Bloxwich. Two years later a supplementary act allowed the extension to Brownhills and thereafter through to Huddlesford on the Coventry Canal. William Pitt was engaged as the engineer and the main line was officially opened on 8 May 1797. This was a boon for the boats and coalfields for it allowed cheap coal to be transported to Derby and Burton, and even as far afield as London.

Water supply was soon a problem and the Chasewater Reservoir was created from the dammed Crane Brook as a solution. The route chosen took advantage of the natural contours in order to minimise the number of locks required – less locks means a quicker journey time. However, this resulted in a highly tortuous route and has given the canal the delightful nickname of the 'Curly Wyrley'. Extensions to the original route fed Hatherton, Cannock, Daw End, Rushall, Bentley and Walsall, while the Anglesey Branch is named from the Pagets, Marquises of Anglesey.

Continued water shortages plagued the system well into the twentieth century, while subsidence from the very mines the canals were created to serve became increasingly troublesome. Indeed the Cannock Extension saw a drop of over 21ft in a single week in the 1940s. Improvements to the trunk roads paid little heed to maintaining an abandoned canal system and much of the route has been infilled. With the demise of the coalfields, the canal and many local businesses served by both, the canal route has been a haven for wildlife or prime land for redevelopment as industrial estates.

In recent years the Lichfield and Hatherton Canals Restoration Trust Ltd have continued the long and daunting target of reopening this route for the leisure industry, not to mention the benefits to wildlife. Today the route is even more daunting an engineering project than it would have been for William Pitt, for today there is the little matter of crossing four lanes of the A38 and A5. Users of the M6 toll road must regard the bridge spanning the six lanes of the motorway with some amusement, for it has nothing connected to either end – at least not yet. This will one day carry walkers, joggers, cyclists, fishermen, fish, wildlife and narrowboats at a top speed of 4 mph (not to mention many thousands of gallons of water) high over the six lanes of traffic racing below.

Featherbed Lane was once a common name for a road which was muddied and waterlogged through continual use by traffic and thus made little or no sound. Fern Croft, Ferndale Croft, Ferndale Close and Ferndale Road all refer to Ferndale, the name of the home of the Round family. Field Road was indeed cut on farmland, Forge Lane and Smithy Lane were the sites of a blacksmith's forge, and Freeford Gardens led to the toll-free river crossing of Darnford Brook. Frog Lane was once Frogmere Street and an area of marshland.

The Garth is a reference to a small enclosure of a yard or garden belonging to a house. George Lane was not known as such until 1730, thus it most likely refers to the reigning monarch. Grange Lane is built on the site of the grange. Hayes View affords views of the region known as Maple Hayes. Hermes Road is after a mythical god which, among other things, is associated with commerce and a perfect option for a business estate. Further, if you want to attract tenants suggest that their prospects may be enhanced by the street name Prospect Drive. Titan Way is named after the Greek god, the equivalent of the Roman god Mercury. It is a name often chosen for industrial areas, for it suggests strength and power. Vulcan Road is after the Roman god of fire which is also called Mulciber in Roman mythology, meaning 'smelter'. These connections with early metalworking are the reason for it being chosen for the industrial estate. Britannia Way is a name that was chosen in twentieth century to illustrate national pride. Europa Way, similarly, was chosen to show modern trade links between Britain and the European community.

Gaiafield Road is an ancient name whose origins are uncertain, for the age very much dictates the meaning, but it has been suggested this refers to the primal goddess Gaia, personification of Mother Earth in Greek mythology. One of the similarly named roads probably gave rise to Gaialands Crescent, Gaia Lane, Gaiastowe and others, and the original is certainly one of the oldest roads in the city. The Leasowe comes from a field name, Lady Leasowe, one of the boundary markers on the Sheriff's Ride of 1805. The field name is a common one, telling of rough pasture. Leomansley Close, Leomansley Court, Leomansley View and Leomansley Road all refer to a minor place name from the Celtic *lemo* and Old English *mansley*, meaning 'elm wood partly cleared for farming'.

Leyfields is recorded since 1298 and is the name of a cottage on the property on a map of 1902. Long Bridge Road speaks of a long wooden causeway, built here to allow horse-drawn traffic to enter the city across almost permanently boggy wetland. Little Barrow Walk is a pedestrian way to the homes at the rear of the Little Barrow Hotel. Lloyds Walk is a pedestrian route created on land given by the owners, Lloyds Bank. Lombard Gardens and Lombard Street was Lumber Street in the early seventeenth century, a name which refers to discarded odds and ends. It took on its present name in 1707, analogous to the London street of the same name. London Road is first recorded in 1245, and the road towards Weeford is Lichfield's example of a name found in every town and city in the land. Sandford Street and Lower Sandford

The old lane associated with Lichfield Cathedral.

Street are recorded as being a sandy track and led to the ford on the Trunkfield Brook, the lower part of the road.

Mallicot Close is named from Mallicot Lane Field, The Maltings has been constructed around the site of the old Lichfield Brewery, and Marsh Lane has been cut on what was known as Marsh Common in the nineteenth century. Market Street cannot have been known as such before 1766 as this was when the market was first held here. Maxtock Avenue is named after a farm on this site in the seventeenth century; Maybank Close is taken from the hawthorn, abundant on this site before development, also known as the May, for this is when it traditionally blossoms; Meadowbrook Road is cut on a meadow which has a tiny brook running through it; Mill Croft wandered a little to its present location, it is a traditional Lichfield name of a path which ran alongside Stowe Pool; The Millpond is where there was indeed a mill pond, one which fed Pones Mill; Miller Close is where the man lived; and Ponesfield Road and Ponesgreen are derived from Pouns Mill, owned by Gilbert Poun in 1180.

Netherbridge Avenue is a field name from maps of 1848 and it refers to the lower bridge. Netherstowe Lane and Netherstowe take their name from Stowe, as do Stowe Croft, Stowe Hill Gardens, Stowe Street and Stowe Road. The Windings is all that remains of a network of old paths which wound their way from Stowe Pool. Nursery Croft was built on land once occupied by a large nursery. Highcroft Crescent and Highcroft Gardens are cut on the grounds of the former house here, which was known as Highcroft. Similarly in the case of Hill Crest Drive the house was named Hillcrest. Hillside was the name of the house and is also the name of the modern street. Cathedral Walk is nowhere near the cathedral, but it is aligned with this famous building and does offer good views.

Pinfold Road marks the site of the pinfold, where stray animals were held waiting collection and payment of the fine. Ploughmans Walk is a clear farming reference, while Poolfield Road takes its name from Poolfield Meadow. Rocklands Crescent is a name taken from a house here. Roman Way is a reminder of the Rykneild Street, a famous Roman road which has somehow become Ryknild Street in modern Lichfield. Rotten Row is a common name from medieval England, always describing poor quality be it buildings or the roads.

St Johns Close, Upper St Johns Street, Lower St Johns Street and St John Street (which has lost its possessive 's') all refer to the Hospice of St John the Baptist. Sandfield Close and Sandfield Meadow both refer to the last remnant of grazing land to have existed within the city boundaries. Saxon Walk is a field name too, from Saxon Nook, but appears only in the nineteenth century and could well be a family name. Shortbutts Lane is taken from a field where crops were grown, here on a record dated 1438. Strips of land, known as butts, were leased to provide produce for the family, with a percentage of the crop being taken in payment. Sometimes the shape of the field meant these strips could not be in a straight line, they bent at right angles to literally produce 'short butts'. The Simpson family gave their name to Simpson Road; Charles Simpson was the fourth generation of his family to serve as town clerk.

Southern Cross marks the point where two roads crossed and could well have been a local name before an official one. Station Road was the original path to the station of Lichfield City, Station Approach is the road to Lichfield Trent Valley Station and here is also found Trent Valley Road. Valley Lane leads to the valley of the Trent. Swan Mews and Swan Road are named after the Swan public house, which is one of Lichfield's leading coaching inns and is first mentioned in 1364. Swinfen Broun Road is named in memory of Colonel Swinfen Broun of Swinfen Hall, whose generous will created the trust named in his honour. Tanyard reminds us of a tannery, once a very necessary part of life although its awful fumes were most unwelcome. Tudor Row is named after Tudor House. Wade Street was once very wet land, as indeed was Watery Lane. Wellington Place recalls the Duke of Wellington public house which once stood here, and Wheel Lane is named after the Wheel Inn. Whitehall is the name of the house and the street; the building was once the Coach & Horses.

Wissage Court, Wissage Croft, Wissage Lane and Wissage Road all come from an old place name which has been so corrupted over the centuries that it is completely unrecognisable and its meaning is unknown. The Woodlands was developed in a predominantly wooded area, as was Woods Croft and Wood Ridings. Weston Road comes from a family of Lichfield lawyers from the mid-sixteenth century, and some later became MPs, registrars of the diocese and chancellors of both Lichfield and Exeter cathedrals. Wightman Close recalls that on 11 April 1611, in Lichfield Market Square, Edward Wightman was the last person in England to be burned at the stake for the crime of heresy. Wilmott Close marks the rise of a Lichfield grammar schoolboy who used his education to make great contributions to the law and eventually became known as Sir John Eardley Wilmott. Langton Court is named after Walter Langton, former bishop of Coventry and Lichfield whose monument is in Lichfield Cathedral.

Field names have given us Austin Cote Lane, Barnfield Close, Beechfield Rise, Beecroft Avenue, Crossfield Road, The Dell, Ellsmore Meadow, Friday Acre, Giffords

Croft, Gorse Lane, Gorsty Bank, Greencroft, Hartslade, Haymoor, High Grange, Hobs Road, Little Grange, Oakenfield, Oakhurst, Orchard Close, Ormonds Close, The Paddock, Partridge Croft, Pipers Croft, Swallow Croft, Thistley Nook, Townfields, Walkers Croft, and Walnut Walk. A theme of tree names has produced Alder Close, Ash Grove, Aspen Close, Birch Grove, Birchwood Road, Cedar Close, Coppice Green, Larch Close, Laurel Close, Lime Grove, Maple Court, Maple Grove, Mulberry Drive, The Pines, Rowan Close, Walnut Grove and Woodfields Drive. Scotch Orchard is another possible field name, although records do not confirm this; however, the later addition of Braeburn Close is a direct reference to it, reflecting both the apple and the Scottish theme.

Road names often tell of their destination, and are given here with the meaning of the place name. Birmingham Road means leading to 'the homestead of the family or followers of a man called Boerma'. Chesterfield Road is a small area near Wall and not the Derbyshire town, yet both mean 'Roman stronghold by the open land'. Curborough Road is the route to 'the mound by the mill stream', and Stafford Road is that which leads to the county town or 'the ford by the landing place'. Tamworth Road and Tamworth Street are named after 'the enclosure on the River Tame', Walsall Road after 'the land of the Welshmen' (which to the Saxons was simply a foreigner), and Whittington Common Road means leading to 'the estate associated with a man called Hwita'. Needwood Hill is a reminder that the Needwood Forest, the 'wood of the poor or needy', once reached as far as the north of the city, while Over Borrowcop crosses Borrowcop Hill or 'the fortified (place at) the hill'. Loosely related to this is Cross in Hand Lane, which remembers the fifteenth-century signpost. This sent travellers in the direction of London, Chester or Stafford – depending upon which direction they were taking.

Individuals who have contributed to the naming of streets include Thomas de Abbenhall who, in the thirteenth century, lived in the region now cut through by Abnalls Lane and Abnalls Croft. Anglesey Road remembers the Paget family, Marquises of Anglesey, as does Paget Close. Anson Avenue brings the family name of the Earls of Lichfield to our attention. The Andrews family owned four of the pubs in Bird Street during the 1960 and are marked by Andrews House.

Benson Close is named after Godfrey Benson, Mayor of Lichfield in 1909 and author of a biography of Abraham Lincoln. Bloomfield Crescent remembers Sir Thomas Blomefield (1849-1928), a justice of the peace who lived in a windmill near Lichfield. Windmill Close and Windmill Lane are named after his home.

Brownsfield Road is a lane leading to land once owned by the Brown family. Bulldog Lane is an old name not adopted until 1974 when it was revived by developers. Charnwood Close is at the rear of Stowe House, home of Lord Charnwood. Erasmus Way and Darwin Close commemorate Doctor Erasmus Darwin, an ancestor of Charles Darwin.

Greenhough Road occupies the property of Richard Greenhough, a maltster in the city. Gresley Row remembers Sir Nigel Gresley of the Drakelow and not the railway engineer, whose only other claim to fame was an unsuccessful campaign to be elected as an MP in 1799. Jackson Road is named after the author of *History of the City and Cathedral of Lichfield*. Harrington Walk is named after Tony Harrington, one of the last teachers at Beacon Street School, as was Bernard Patrick who gave his name to Patrick Mews. Harwood Road remembers the Reverend Thomas Harwood who

Lichfield's cathedral dedicated to St Chad.

served as headmaster of the grammar school from 1791 to 1813 during which time he wrote a history of the city from the viewpoint of the church. Caterbanck Way is named after a family who have had ties with Lichfield since at least the seventeenth century, when one Thomas Caterbanck married Anne Mitchell in Lichfield Cathedral, 7 October 1677.

Occasionally we find spelling errors, Hayworth Road being one example as Ideally it would appear the same as the man it was named after, Bishop Heyworth, Founder of the almshouse in Beacon Street in 1424. Hewitt Close remembers John Hewitt (1807-78), a principal figure in the War Office who organised much of what we still see today in the armoury at the Tower of London. His love of literature led to him writing extensively on the subject, under the extraordinary pseudonym Sylvannus Swanquill. King's Hill Road is a reminder this is the site of an ancient burial ground of Saxon nobility. Knowle Lane is a part of the former Roman road, which took the name of an eighteenth-century farmhouse called Knoll – telling us it was on a small rounded hill.

Colonel Longstaff was Sheriff of Lichfield in 1918, and Longstaff Close was developed on the Nearfield Estate where he resided. In 1654 William Lunn left two houses, at what is now Lunn's Croft, to be used as almshouses for six poor widows. When the houses were demolished and modern homes erected the benefactor was commemorated. Lyn Avenue and Lynfield Road recall how, in the early nineteenth century, pianist and composer Muzio Clementi lived here in the house and estate called Lyncroft.

Lewis Close remembers how Joyce Lewis of Mancetter was burned at the stake in 1557 for heresy, the only female martyr of these religious troubles. Lomax Close remembers Alfred Lomax, the son of the founder of a noted local publisher of religious books. Lillington Close tells a little of the history of Lichfield. In 1705 a meeting at the Kings Head in Lichfield raised a regiment of regular infantry, which eventually became the 1st South Staffordshire Regiment. The principal figure in these early days was Colonel Luke Lillingston, so while it is a fitting reminder of the man it is a shame they did not spell his name correctly.

Manley Road reminds us of the family resident at Manley Hall and who held lands here. Martin Croft remembers how George Martin (*d.* 1908) provided funds for Lichfield Victoria Nursing Home. When it opened in 1899 he gave houses for use as both offices and a home for the nurses. Maxwell Close is named from a family named Maxwell living here in 1870, later transferred to the land as Maxwell's Place. Mesnes Green is a name which has appeared as Maines, Menzies and Mains over the years, the council eventually settling on the oldest spelling; it is undoubtedly a family name, yet which family is unknown. Minors Hill comes from Thomas Minor, a mercer of Lichfield who served as MP 1654 to 1660 and who founded a school which survived until 1876.

Nearfield Hill remembers the house formerly owned by Colonel Longstaff, and an old map shows two fields: Daniels Near Field and Daniels Far Field, hence the name. Oakley Close is named after the Sir Charles Oakley, a former governor of Madras who lived here. His sixth son, the Reverend Frederick Oakley, was the author of the popular carol *O Come All Ye Faithful*. Queen Street was cut in the same year as the coronation of Queen Victoria. Quonian's Lane's historical records of this name show a bemusing number of different spellings, but the suggestion it comes from

The modern name comes from a corruption of Conyngham, meaning 'king's house'.

Latin *quoniam* (because/whereas) is highly unlikely. The best offering is that it is a corruption of the family of Conyngham, who took their name from the house here – Conyngham means 'king's house'. The area was recorded as Kings Meadow Land in a document from the late twelfth century.

Reeve Lane shows a title which has existed since Saxon times, the reeve was appointed by the lord of the manor to see to the running of the estate. Romilly Close recalls the law reformer Sir Samuel Romilly and his son Baron John Romilly, neither of whom have any direct link with Lichfield. John Romilly, however, was a judge at Warwick in March 1862 at the famous legal case of the Swinfen family. This was one of four incidents at the beginning of the decade which led to the regulation of barristers in the country. Prior to this barristers were not bound by anything other than moral and traditional obligation, by the end of the decade a strict set of rules was in force.

Seckham Road was occupied by a family of this name when the road was first cut. Shaw Lane could be a place name but the late appearance may indicate a family name is more likely. The famous Sheriff's Ride of Lichfield was directed along what is now Sheriffs Close in 1976. Spearhill is a field name recorded in the sixteenth century, possibly where poles were cut from coppiced or pollarded woodland as shafts for spears. Sturgeon's Hill is probably taken from a David Sturgeon who is recorded as the owner of property near here in the middle of the eighteenth century. Stychbrook Gardens is from the Stych Brook and Farm and the name speaks of this being a drainage channel.

Victoria Gardens was chosen to complement the Victoria Hospital, Victoria Square because the Victoria Barracks occupied this site in 1854. Wittell Road was built on the site of the old Lichfield Brewery, the company motto of which was 'Quality Will Tell'. The company name was transferred to that of Wiltell Works, and from there the name was transferred to the road. Yew Tree Avenue is named after Yew Tree House, which was once a public house.

Several roads remember the names of senior bailiffs, the equal of the mayor before the seventeenth century: Allington Avenue remembers John Allington; Bardell Close, pewterer Edward Bardell of Wade Street; Chatterton Avenue, John Chatterton; Collins Hill, George Collins; and Deykin Road, mercer Nicholas Dreykin. John Fecknam also left money for a hospital and well merits the naming of Fecknam Way. Newbould Close is named after a mercer, Richard Newbold, also served as sheriff in 1625, while Tewe Place is named after fishmonger Humphrey Tewe. Levett's Square is named after one of the most influential of families in the history of Lichfield and surrounding areas. Other Lichfield officers have given their names to streets, including those who are commemorated by Burwaye Close, Heathcot Place, Maddocke Walk, Stonnyland Drive, Stonynge Place, Thropp Close and The Whytmore.

Arguably Lichfield's most famous son, Dr Samuel Johnson's influence is not restricted to the pub which bears his name, or even Samuel Close and Johnson Close. Bolt Court is named after the place off Fleet Street in London where the man died in 1784. Garrick Close. Garrick Court and Garrick Road come from David Garrick, pupil of Johnson and the greatest actor of his time. Gilbert Road is in memory of Alderman John Gilbert, a benefactor of the city who enabled the purchase of Lichfield, Samuel Johnson's birthplace. In his first dictionary Samuel Johnson dedicated the milestone work to his native city, in return Lichfield's coat of arms quotes that dedication as its motto *Salve, magna parens* – 'Hail great Mother'.

Hawkins Close remembers Sir John Hawkins (1719-89), a member of the famous 'club' founded by Doctor Samuel Johnson. He published *Life and Works of Johnson* (eleven volumes) and was named as an executor of Johnson's will. The Parchments is a reference to the parchment works, owned by bookseller Michael Johnson, Samuel's father. Willow Tree Close is derived from Johnson's Willow, the willow tree on the North Bank of Stowe Pool, a place close to the site of Michael Johnson's parchment factory. The original willow was visited by Dr Johnson, but it blew down in 1829. A second tree grown from the first lasted until 1881, and then a third generation cutting was grown. It was removed in 1956 when a fourth cutting had been rooted and planted there.

Hunter Close recalls the man who was not only Samuel Johnson's headmaster but also grandfather of Anna Seward (1747-1809), the celebrated 'Swan of Lichfield' who is honoured by Seward Close. In her time she was a leading figure of Lichfield society, and a friend of Sir Walter Scott and Sarah Siddons. Interestingly, letters of the day reveal how Dr Johnson and Miss Seward shared a mutual dislike for one another.

The life and work of Sir Walter Scott (1771-1832) forms the basis for a theme of some street names in Lichfield, deservedly so for he could be said to have been the first writer in the English language to have enjoyed universal appeal (where the English language was understood). In the twenty-first century his works are still seen as classics. They include *Ivanhoe, Rob Roy, The Lady of the Lake, Waverley*, *The Heart of Midlothian* and *The Bride of Lammermoor*.

Sir Walter Scott, who inspired the naming of Scott Road, suffered from polio at the age of two. This left him lame and resulted in him going to live with an aunt who taught him to read and write. It was while living in the Border region at this young age that he learned the legends and stories that gave his work a unique flavour. He followed his father in the legal profession and did not start writing until the age of twenty-five. His first novel *Waverley*, after which Waverley Walk is named, was initially published anonymously in 1814 and released as an attempt to offset personal financial difficulties.

The subject matter – that of the Jacobite rising – proved most popular in his native Scotland and was the first of several novels set in his native Scotland over the next five years. All of these were published under the pseudonym Author of Waverely so as not to detract from his reputation as a serious poet. One of this series was entitled *Kenilworth*, published in 1821, and this led to the naming of Kenilworth Road. At this time he was nicknamed the Wizard of the North.

In 1819 his writings broke away from Scotland with *Ivanhoe*, the first of a number of novels set along the same lines, which inspired the naming of Ivanhoe Road. Six years later, with his company again in financial difficulties, his identity became known officially and his novels gained a new lease of life; Sir Walter Scott proceeded to write his way out of debt.

He was among the most prolific of novelists with most of his work being completed in the last seven years of his life. He died at his Abbotsford estate, which gave a name to Abbotsford Road. A gauge as to his standing in the literary world can be judged by a complete lack of punctuation in his manuscripts, leaving it entirely in the hands of the publisher – which, as this author knows only too well, is not to be recommended today.

There are a number of streets marking the battles of the Civil War, of which Lichfield saw more than its fair share. Brook Close commemorates Lord Brooke, who captured Lichfield for the Parliamentarians in March 1643. It lost its final 'e' in an official document of the 1960s. Bunkers Hill refers to where cannons were housed to limit movement when being fired. Cromwells Meadow is said to have been where the Parliamentarians camped during one of the sieges of the city. Prince Ruperts Mews and Prince Ruperts Way both refer to the location of this commander's artillery during the siege of the city. The Napoleonic Wars are also remembered. Being far inland it was the ideal place for French prisoners of war. Frenchmans Walk reminds us that, enemy or not, some were given freedom and even educated local children in their native tongue.

Angorfa Close takes its name from Angorfa House, which was here until the 1960s. Home to the Heath family for many years, nothing has ever been traced regarding the origin of the house name. Laburnum Court is for once not a creative name but a street where laburnum trees were plentiful. Lichfield's twin towns have provided the names of Saint Foy Avenue and Limburg Avenue, towns in France and Germany respectively.

Any suggestions of Cricket Lane being named after the insect are only due to an origin from the sport yet to be proven. The origin of cricket in Lichfield is uncertain, but the assumption is that it started on the site near the Shoulder of Mutton, hence the name of the street Kensington Oval, which is known to be the venue for horse and pony racing.

Lichfield Cricket Club has provided some street names. Harman Drive remembers Peter Harman, a serving police officer and talented cricketer. He was sadly to lose his life in a road traffic accident. Lawrence Way is after William Lawrence, whose loyalty to the club saw him serve as a player, umpire, scorer, secretary and committee member. Other names derived from the history of cricket and hockey in Lichfield include Bains Drive, Blakeman Way, Broadbent Close, Colling Drive, Lightwood Close and Whitehouse Drive.

Famous racecourses gave names to where horses were trained and resulted in Ascot Close, Epsom Close, Goodwood Close, Lambourne Close (despite the addition of an erroneous 'e'). Sunbury Avenue is named after a town in Surrey, home to Kempton Park racecourse, and Saddlers Close is a less direct reference to the sport of kings. Wentworth Drive is named after the family of Ann Wentworth, Countess of Strafford, who, along with other titled enthusiasts, was in almost permanent residence during the annual Lichfield race weeks.

National figures are commemorated too, sometimes directly and sometimes indirectly. Ashmole Close and Elias Close are after Elias Ashmole, founder of the Ashmolean Library and Museum at Oxford and who was born in Lichfield; Auchinleck Drive is named after the Scottish home of James Boswell; Smallridge is named after George Smallridge, who became Professor of Divinity at Oxford and Bishop of Bristol and was a pupil of Elias Ashmole. Bloomsbury Way is named after the literary group who lived in or near London in the first half of the twentieth century. A controversial and influential movement, the criteria for membership was never stated!

Eastern Avenue is a modern by-pass, designed to direct traffic to the east, Western By-Pass its mirror equivalent. Quarry Hills Lane supplied stone from the fourteenth to the nineteenth centuries, and it was used for the bishop's palace and the cathedral.

Theme names are popular for modern developments. Drake Croft and Mallard Croft are a pair of streets named after ducks. Newton Road is named after two Lichfield brothers: Andrew Newton, who funded the building of houses later to become known as Newton's College and Thomas Newton, who went on to become Bishop of Bristol. Newtons College is also a Lichfield street name. Wolsey Road is named after Cardinal Thomas Wolsey (1475-1530), while Heenan Grove is certainly after Cardinal John Heenan (1905-75), Archbishop of Westminster. The naming of these seem quite abitrary, yet it could have been deliberate to balance with streets named after Church of England bishops.

On the All Saints Estate the street names were, predictably, chosen to reflect the name of the estate. Hence we find Augustines Walk, Christophers Walk, Francis Road, Giles Road, Jude Walk, Lukes Walk, Marks Walk, Matthews Walk, Pauls Walk, Peters Walk, St Annes Road, St Catherines Road, St Chads Close, St Chads Court, St Chads Road, St Helens Road, St Margarets Road, St Marys Road and Stephens Walk. Thomas Greenway and James Greenway also refer to saints; the addition 'Greenway' is not a surname but a modern term for a pedestrian right of way.

Streets with a royal connection are Balmoral Close, named after the official Scottish residence of the royal family; Broadlands Rise, the home of Lord Mountbatten; and Buckingham Gardens, the official residence of the reigning monarch since 1837. Grosvenor Close was chosen to honour the family name of the Dukes of Westminster, allied to royal names hereabouts. Marlborough Croft refers to the family of Sir Winston Churchill, the Dukes of Marlborough. Richmond Drive is a name suggested by residents to complement other streets referring to the nobility of the land. However, while Richmond may have a long history, there is no title.

St Chad's Cathedral and those connected with it gave names to Cathedral Rise, Chadeswell Heights, The Close, Deans Croft, Deans Slade Drive, Friars Alley, The Friary, Friary Avenue, Friary Gardens, Friary Road and Minsterpool Walk. Pennys Croft takes its name from Dean Penny, who was at the cathedral in 1750. Vale Close refers to the spires of the cathedral being known as the Ladies of the Vale. Vicar's Close is a part of the cathedral's outbuildings, erected to house junior members of the church. After this the cathedral would have been the inspiration for a theme of other cathedral cities leading to Canterbury Close, Chester Close, Gloucester Close, Lincoln Close, Norwich Close, Salisbury Close, Southwark Close, Truro Close, Worcester Close, Winchester Close and York Close (correctly a minster).

One of Lichfield's oldest churches.

A literary theme has produced Burns Close, after poet Robert 'Rabbie' Burns; Byron Avenue, poet Lord Byron; Chaucer Close, Geoffrey Chaucer; Masefield Close, John Masefield; Shakespeare Avenue, William Shakespeare; Spencer Road, poet Edmund Spenser (although the spelling is wrong); Stevenson Walk, Robert Louis Stevenson; and Wordsworth Close, poet William Wordsworth. Composers were the theme for Elgar Close, Handel Walk, Purcell Avenue, Sullivan Way, Sullivan Walk, and Verdi Court. Reynolds Close is named after Sir Joshua Reynolds (1723-92), seen as England's greatest portrait painter. Thespians have provided another theme, actors and actress to be so honoured include: Irving Close (Sir Henry Irving), Kean Close (Edmund Kean), Jordan Close (Dorothea Jordan), Siddons Close (Sarah Siddons) and Terry Close (Dame Ellen Terry).

One street name theme is that of Staffordshire Regiment members awarded the Victoria Cross. A theme not immediately apparent, owing to the streets being scattered and not confined to a specific area. Baskeyfield Close remembers Sergeant John Baskeyfield, honoured posthumously in 1944. Booth Close remembers the instructor at Whittington Barracks who fought in the Zulu Wars. Carmichael Close is for Sergeant John Carmichael, killed at the 3rd Battle of Ypres. Coltman Close is after Lance Corporal William Coltman, a stretcher-bearer in the First World War. Flinn Close remembers one T. Flinn, an Irish drummer boy during the Indian

The old friary reminders are still found.

Mutiny, when he took on three mutineers single-handed, and all when he was just fifteen years of age. Henderson Close reminds us of the bravery of Lieutenant Colonel E. A. D. Henderson who led his men in repeated attacks against Turkish troops at the Battle of Kat-Al-Amara in 1917. Wounded in the encounter, he died shortly afterwards.

When choosing a theme it is probably only in hindsight that some subjects should have been avoided as there is a finite number of names available. The seasons of the year being one example, which only yielded Autumn Drive, Spring Road, Summer Grove, and Winter Close. Similarly were it not for prior knowledge the theme linking Holywell Rise, Mawgan Drive, Newlyn Close, Pentire Road, Tregony Rise and Trenance Close would have been unknown, for they were suggested by a director of the developer as a reminder of his numerous family holidays in Cornwall. Lawford Avenue is a name given by developers from an architectural style seen around Church Lawford and Long Lawford near Rugby.

Street names stemming from modern developments, and which have no local etymological value, include Bluebird Close, Cornfield Drive, Curlew Close, Hawksmoor Drive, Hazel Grove, Keepers Close and Park End. The developers themselves always offer a name for the project which can be retained for the name of the street, and thus the following were created: Hawkesmoor Close, Hawksmoor Drive. Heathcote Drive, Heather Close, Henley Court, Lanthorn Close, Stonnyland Drive, Wharf Close, Willow Court, Windsor Court, Witley Drive, Aiden Court, Aiden Court Road, Barn Close, Blackthorne Road, Bracken Close, The Brambles, Copperfields, Cranleigh Way, Fallow Field, Gable Croft, Gledhill Park, Havefield Avenue, Hawthorn Close, The Hedgerows, Heritage Court, Manor Rise, Maryvale Court, Orchard Grange, Shepherd Close, Sorrel Close, The Spires (five are visible from here), The Sycamores, Tudor Close, Warren Close and Willowsmere Drive.

## Little Wyrley

Wyrley Lane is clearly a destination name, the place name meaning 'the woodland clearing where bog myrtle grows', and Fishley Lane too heads south to the neighbouring Bloxwich and is a minor place name. Gorsey Lane refers to the vegetation which has long been abundant here. Grove Lane marks the location of Wyrley Grove, the ancient seat of the Fowkes family. It was later the residence of the Hussey family, who came to the hall when heiress Sybil Fowke married Joseph Hussey of London. Later, in 1834, the mansion was owned by John Stokes. School Lane is of obvious derivation, although there is no school here.

## Lynn

Lynn Lane is clearly a destination name, albeit leading into the village. Mill Lane refers to the local mill, and Thorneyhurst Lane features a quite common landscape name.

# Chapter Nine

# Mile Oak—Muckley Corner

## Mile Oak

Standing at a major crossroads it is not surprising that the street names include destination names of Bonehill Road, Hints Road and Sutton Road.

Field names include that of Plantation Lane, which, had it not extended over the years, would not even have reached as far as Mile Oak. Brookside Road runs up to the Bourne Brook. The Green was the site of the original village green.

Close proximity to the Drayton Estate gave names to both Manor Court and Manor Road, while Oak Drive was developed on land where the former Mile Oak Hotel stood.

Price Avenue remembers how Sammy Price bought Bonehill House in the 1930s. F.G. Allton was the owner of a business in the 1920s. He lived at Bonehill Lodge, was a baker by trade, owned a number of shops in the area, worked as a local councillor and served on Fazeley Borough Council. Allton Avenue justly remembers this man.

Heathcote Drive honours Councillor Mrs P.M. Heathcote, a member of the parish council in 1963. Two members of the Tamworth Rural District Council are honoured in Affleck Avenue and French Avenue, respectively Mr R. Affleck and Mr W. J. French. Brewster Close is named after Mr R. C. Brewster, a Trustee of the Mile Oak Sports and Social Club when it was founded 25 March 1945. George Avenue remembers the coronation of George VI in 1937, while the coronation of Elizabeth II in 1953 coincided with the development of Coronation Avenue.

Cranwell Rise is a misnomer. It is built on land which was granted by the priory of Canwell for a mill, and from which it is possible to look towards, although not actually see, the region of Canwell to the south on the road to Sutton Coldfield. How the name became corrupted as Cranwell from Canwell remains a mystery.

The reasons for creating the names of Aldin Close, Caistor Close, Gainsborough Drive, Kirkland Way and West Drive are not documented, making it impossible to say where they originated.

The land owned by Canwell Priory has been remembered incorrectly.

Which came first – Boat Lane or the Boat Inn opposite, near Muckley Corner?

## Muckley Corner

Muckley Corner undoubtedly comes from Muckley Field, a common divided up in the nineteenth century. Coppice Lane takes its name from the coppice, a small wood or thicket, long gone. Cranebrook Lane is taken from the Crane Brook. Boat Lane recalls the former canal here, although it is debatable if the road was named before the Boat Inn, which was previously known as the Old Boat Inn. However, the road was certainly here before the pub, and the reason for its name on the Essington and Wyrley canal is clear, it is simply the date the name came about which is uncertain.

Hall Lane, a destination derived street name, serves Hammerwich and Hammerwich Hall, as does the Walsall Road. Watling Street runs diagonally across the country and includes St Albans in Hertfordshire. The Saxons knew the place as Waetlinceaster 'the Roman stronghold of the Waetlinc tribe', and this is the origin of Watling Street.

# Chapter Ten

# Norton Canes

Watling Street can hardly be considered to have a name derived from local history. Indeed, the name is derived from a tribe living around the Hertfordshire town of St Albans called the Waeclings. Chasewater Way is named from the nearby stretch of water formerly known as Norton Pool, hence Pool Avenue also. Norton Grange Crescent, Norton Grange Lane, and Norton Hall Lane mark the position of important houses, while Park Road marks the grounds. Norton Springs is clearly a water supply, Long Lane lives up to its name, Pinfold Lane is an old name where stray animals were impounded and Lock Keepers Close was the home of the man who looked after the lock on the canal. Church Road and Church Vale both lead to, and end at, the site of the church dedicated to St James the Great and similarly named is Chapel Street. However, St James Road, almost certainly named for the same church dedication, is quite far away from the church itself and the connection between the two is unknown.

Bridges Crescent is derived from Bridges Road, itself a reference to a person rather than a bridge. Red Lion Lane clearly features the most popular public house name in the land – at its peak there were over 600 examples, although this has diminished in recent times. This was, as with all coloured animal names of pubs, heraldic in origin and most often referred either to John of Gaunt or Scotland. Hanbury Road recalls Robert William Hanbury, who established the Coppice Colliery in Heath Hayes.

Jerome Close and Jerome Drive are both derived from Jerome Road, and are ultimately a reference to author Jerome K. Jerome. School Close, School Crescent and School Road are all derived from Norton Canes Primary School which is to be found here. Breeze Avenue reminds us of blacksmith Richard Breeze, here at the end of the nineteenth century. Hussey Road is named from the family resident at Wyrley Hall from at least the seventeenth century, and who were influential landowners in this area.

Butts Close and Butts Way share a common origin in the term describing the strips of land at each end of a field which are not ploughed but are where the plough and the team pulling it turn. High Street is the name given to the most important street in the early days of a settlement. Hill Street is most aptly named, albeit a short climb here, although Lakeside Drive is not alongside the stretch of water but facing it.

One major theme around here is that of horse racing. This is seen in the name of Newmarket Road, which refers to a Suffolk racecourse known as the Home of Racing, not only in Britain but the world. Featuring two courses, the Museum of Racing, the National Stud and the centre for stock-breeding and equine medical treatment. Newmarket was founded literally as the 'new market' by Sir Richard de Argentein in 1200. In 1605 James I brought sport to Newmarket in the form of hare coursing. It is said that this was when the first horse races took place, the palace courtiers were racing Spanish horses, descendants of the animals brought as part of the invading Spanish Armada of 1588, which had come ashore in Galloway. Charles I (reigned 1625-49) had the first grandstand built here, and at the Restoration of the Monarchy in 1660 Charles II firmly rooted the sport here by constructing Palace House nearby in order to accommodate the royal court on its annual sojourn to Suffolk. It was then that Town Plate was established, the oldest horse race in Britain run under written rules. That race is still run every October, on the July course. The other course is called the Rowley Mile, named after Charles II's favourite horse Old Rowley, a nickname also given to the king himself.

Other street names here are Lingfield Road, from the Surrey course of Lingfield Park; Newbury Road, a Berkshire course where racing began in 1805; and Sedgefield Way, which reminds us of the National Hunt course at the southern end of County Durham. Lambourne Way – which is not exactly a course name, but a reference to the Valley of Lambourn and the Berkshire downs where horses have been trained for many years – is one of three names which refer indirectly to racing, another being that of Turf Close. The final name of Beaumont Way is less certain, it is possibly that of experienced and well-known racehorse trainer Peter Beaumont, although this would be an odd diversion from the others. There is a Beaumont Park course, however, that is on the island of St Kitts and Nevis in the West Indies, unusual for there are certainly many other British racecourses to provide a name.

Another theme is that of trees, which has produced Cherry Tree Road, Elm Road and Silver Birch Road. Destination names include Brownhills Road, Hednesford Road, Norton Lane and even Norton East Road and Norton Green Road are simply routes through the town. Conduit Road is the main thoroughfare on the industrial estate, and Rolling Mill Road recalls an early industry. Railway Street and Railway Walk are of obvious derivation.

In modern times it seems that almost every new street has to be given a name which suggests a more rural location. Hopefully, this trend is a mere passing fad and before long street names will begin to reflect the real history of an area. Otherwise every town and city in the UK will have many identical street names. Developments in Norton Canes have been given names of Arbour Mews, Bellsize Close, The Brambles, Broadoaks Close, Castlecroft, Chequers Court, Cherry Brook, Deer Close, Gainsbrook Crescent, Grange Road, Hamilton Lea, The Hamlet, Hope Drive, Ivy Gardens, Leighswood Close, Mere Croft, Pennycress Green, Quinton Place, Richards Court, Spinney Close, Stag Crescent, Victoria Court, Wallace Close, Woodfield Close, Woodfield Drive and Yew Tree Close.

There are always those names which must have been given for a reason, but that etymology is now lost, and such is the case with Braemar Road, Charter Close, Jackson Close, Knights Court, Legion Close, Poplar Street, Stokes Lane and Washbrook Lane. Without further evidence it would be pure speculation to suggest any origins, even if they seem quite obvious.

# Chapter Eleven

# Packington—Pipehill

## Packington

Jerry's Lane has been named such since at least the beginning of the eighteenth century. It is certain that this refers to a former resident here although there is no one of this name recorded. It does not seem likely to have been anyone at Packington Hall, built by architect James Wyatt in the eighteenth century for the Levett family.

View from the highest point on Cannock Chase.

## Penkridge

Pottal Pool Road is taken from an old pool name, itself most likely a corruption of a name meaning 'pot well'. Broadhurst Green Road is taken from a minor place name, meaning 'the wide wooded hill of or by the green'.

## Pipehill

In the earliest times, Broad Lane would have been quite a lot wider than the other streets here, and Peter's Lane most certainly refers to a former resident, even if the person is not documented. One destination name here is Lichfield Road, leading to 'the open land near Letocetum', the Roman name for Wall.

# Chapter Twelve

# Shenstone—Swinfen

## Shenstone

Main Street is the most important in the community. Church Close stands next to that dedicated to St John the Baptist. New Road was 'new' when it was first developed around 1900. Birmingham Road is the one surviving destination name.

Destination names here include Birmingham Road and Lynn Lane, and Streetway Lane is derived from the name of Watling Street and refers to it being a Roman road. Millbrook Drive, Mill Court and Mill Lane refer, somewhat predictably, to the former mill here and Station Road runs past the railway station. St John's Hill, St John's Drive and Church Road all refer to the church dedicated to St John.

Aston Close takes its name from the ancient place of Aston Parva. Birch Brook Lane means exactly what it seems, the east ridge gave its name to Eastridge Close. Chestnut Drive, Hollyhill Lane and Oakwood Close all reflect the landscaping of Shenstone Court. Fotherley Road is a contraction of the original minor place name of Footherley, which was also taken by Footherley Hall. Holm View Close is also derived from a place name, and Hook Lane is taken from Hookfield. Lincoln Croft comes from Lincoln House, itself probably named from the first family to reside there. Park Lane runs alongside an ancient estate, owned in part by Alexander Ward during the reign of Charles I. What was Shenstone Court gave names to Court Drive, Admiral Park Drive, which remembers a former owner, as does Richard Cooper Road.

Shenstone Court was home to the Baronets Cooper, a title created on 20 December 1905 for Richard Powell Cooper. The family were unheard of before the Victorian era, when the opportunity to rapidly rise in prosperity and status was finally open to all, when they established a family business that provided insecticides and vetinary products, principally the cocktail which was used in a sheep dip to rid the animals of pests quickly and efficiently. One of the family is reported to have claimed to have 'Saved Australia's economy overnight', the island's prosperity being almost entirely based on its enormous sheep farms at the time.

Ashcroft Lane takes its name from a mansion called the Ashcrofts, once the estate of the Brandreth family. Barnes Road reminds us of the family who were here in the eighteenth century. Churchill Road is named after the owners, John and Henry of

St John's Church, Shenstone.

that family name, who were based at Hints. Essington Close remembers long-serving vicar Robert William Essington who served from 1848-95. He retired to Newquay, Cornwall, where he died on 18 July 1907 at the age of eighty-nine. Foden Close honours John Foden, clerk of the parish in the sixteenth century. Pinfold Hill is a reminder of the work of the pinner, who rounded up stray animals and held them until payment of the fine. Raikes Lane is a corruption of Rakemoors, itself describing a rough piece of land. Doctors Lane is as obvious as Schoolfields Road, where the place of education was built in 1865.

## Streethay

The Holland family owned the Streethay estate during the eighteenth, nineteenth and twentieth centuries and gave their name to Holland Close. The Reverend John Bailyes, recorded as living here at Bexmore House in 1790, gave his name to Bailey Close.

Dyott Close is named after one of the most important families in Lichfield's history. They have held Freeford for hundreds of years, and although the line can be traced back to Robert of Freeford around 1260, the surname is not seen until 1563 and John Dyott. This man is thought to have been the basis for the character John Doit of Staffordshire, the companion of Justice Shallow, in William Shakespeare's *Henry IV Part II*. John Dyott, his son Anthony, his wife Dorothy and their son Richard reunited the manor, which had become divided through inheritance by three sisters. From there the manor came back to the Dyott family, several of who were barristers and many represented Lichfield in parliament right through to the nineteenth century.

Brookhay Lane is a local place name that seems to have first appeared as a minor station name on the Walsall to Burton line. The station is recorded in 1847 but had apparently disappeared just three years later. This would logically point to the name being taken from a map prior to 1847, yet when the station was removed shortly afterwards nobody bothered to reinstate the place name – which speaks of 'the brook by the forestry region'. The station was next to the level crossing and probably only ever intended to be a temporary and unofficial stop, ostensibly for the workers on the railway itself. Bexmore Drive is from Bexmore Farm, demolished to make way for Eastern Avenue, and which comes from a twelfth-century place name of *Berksmoor* meaning 'the birch moor'.

Burton Old Road features an old destination road to Burton on Trent, now closed off and replaced by the A38, and Burton Road is the same. Ash Tree Lane has many ash trees, and Meadow Croft speaks of the former 'meadow and crofts'.

## Swinfen

The Swinfen estate has always been confined to just the grounds of the hall here. Indeed the name of The Drive shows it to be the road leading to the hall itself. The only other name of note is Broom Covert Road, a landscape name describing the small wood of broom trees. The hall was originally built for Samuel Swinfen, his wife and fifty-five servants and staff. It completed in 1757 under local architect Benjamin Wyatt.

# Chapter Thirteen

# Tamworth

Tamworth's history is as old as any in Staffordshire and this is reflected in its street names. The oldest are seen in the following: Aldergate or 'Ella's pathway'; Gungate, and Lower Gungate and Upper Gungate (all refer to 'Gumpi's gate or pathway); Alders Lane and its alder trees; Bolebridge Mews, derived from Bolebridge Street, itself speaking of an ancient river crossing; High Street, always historically the main street in a settlement; King Street and Market Street, off which it runs, are named to commemorate the centuries old granting of a market to Tamworth; Ludgate and Lud Lane together describe a cattle track; Moor Street and Moorgate are the roads to the moorland; Colehill used to be known as Cross Street after the Stone Cross which stood near here – now only commemorated by the name of the pub – and although the street name does indeed stand on a hill, the first element has never been understood. The modern Cross Street at The Leys forms an obvious cross on the map with Cherry Street, while the Cross Street at Kettlebrook crosses a number of other roads.

Holloway is the well-worn track south of the town leading across the river, and even today the road is below the level of the footpath and the houses. It is quite literally 'the hollow or well-worn way'. The two Moor Lanes, one at Amington and the other at Bolehall, are at opposite ends of the Warwickshire Moor, which has since become a part of Staffordshire. Salters Lane is an ancient trackway named after men who used it for the transportation of the vital commodity of salt, once the only way to preserve meat. Slade Lane is a common name found among old pathways and fields, referring to land in a marshy valley.

Edgar Close is named after the Mercian King Edgar, and Edward Court and Edward Street, after the Saxon King Edward the Elder. Ethelfleda Road marks the woman who assumed control of Mercia after the death of her husband. Ethelfelda oversaw the rebuilding of Tamworth following the wars with the Danes and died in the town on 13 June 920 when she was over sixty years of age. She is interred in Gloucester Cathedral alongside her husband in the eastern porch of St Peter. Offa Street and Offa Drive are named after the ancient king of Mercia (ruled 757-796), of which Tamworth was once the capital. The building of Offa's Dyke, the defensive feature between England and Wales, is attributed to him. Athelstan Way remembers Athelstan (895-939), the last royal resident in Tamworth, the great nephew of King Alfred the Great.

The Saxon era was marred by skirmishes with the Scandinavians, eventually leading to a large region in the north and east of England coming under their control. This region was known as Danelaw or, as the road is known, Danelagh Close from Old English *dena lagu*. Manor Road does indeed mark the position of a former manor.

In later years, as Tamworth developed as a market town, other residents gave their names to streets. Freville Close is named after a former lord of the manor, Sir Alexander Freville (died 1328). Hastings Close commemorates the family name of the Earls of Pembroke, associated with Tamworth since the fourteenth century. Marmion Street recalls the Norman baron who was here at Tamworth, Sir Robert Marmion (*d.* 1101). Ferrers Road recalls former lord of the manor, Thomas Ferrers. Cromwell Road instantly conjures up the idea of Oliver Cromwell, Lord Protector during the Commonwealth, and yet this same family had branches around Staffordshire who are more likely to have been the inspiration for the name.

Allen Street is either named after John Allen, bailiff of Tamworth in 1592, or a man with the same name who was town clerk in 1664. Dumolos Lane remembers Mr Dumulo of Dunton House, Coleshill towards the end of the nineteenth century. Robinson Close remembers when William Robinson, an influential gentleman of his day, was resident at Bonehill Cottage. Thomas Street is probably named after Richard Thomas, an insurance agent based in Hospital Street at the end of the nineteenth century. Thirty years later Justice of the Peace Thomas Florendine was an important figure in Tamworth, his name being given to Florendine Street and thence to Florendine School.

Brain Street takes the name of George Brain, who lived at 552 Main Road, Glascote in the 1930s. George can certainly lay claim to having one of the oddest reasons for a street name in the town's history. His job was allied to the open cast mining; he was the only man in the town who was licensed to carry explosives.

Street names related to destinations also develop naturally and here we find the following: Amington Road, 'the farmstead associated with Earma'; Ashby Road, leading to Ashby-de-la-Zouch, a place name speaking of 'the settlement near the ash trees belonging to Richard de la Zouch'; Fazeley Road which leads to 'the clearing where bulls are reared'; Glascote Lane and Glascote Road which refer to 'the cottage of hut where glass is made'; Hockley Road, which probably refers to 'Hucca's woodland clearing'; Lichfield Road and Lichfield Street, which were named from the city with a name meaning 'the open land near Letocetum (the Roman name for Wall)'; Bitterscote Lane, which is likely to have been 'the cottage of a man called Bytta'; Bonehill Road, which could refer to either 'Bula's hill' or 'the hill of the bullock', both are equally plausible; Comberford Road, which leads to 'the ford of the Cumbrians'; Wigginton Road, which takes us past 'the farmstead associated with a man called Wicga'; Dosthill Road, a name meaning 'the hill with a shelter for animals'; Kettlebrook Road, which is named after the Kettle Brook, a river name which has never been understood but is thought to be a personal name; Moor Lane, which leads to the former Warwickshire Moor, site of a major annual market; Shuttington Road, a place name meaning 'the farmstead associated with Sceot's people'; Stoneydelph Lane, which is quite literally 'stone diggings'; Tamworth Road (actually more than one), which all lead to 'the enclosure on the River Tame' with the river name referring to 'the dark one'; Wilnecote Lane, once 'the cottages of Wilmund's people'; Whateley Lane, once 'the woodland clearing where wheat is grown'; and finally Belgrave Road, a name describing 'the fine or beautiful grove'.

While not exactly a destination name, the Watling Street is considered one of the country's major Roman roads and its name is derived from the Waetlinc tribe resident near present St Albans. Alluding to the Roman influence are Hadrians Close, after the famous wall, and Appian Close, recalling the Appian Way, the most famous Roman road in Italy. Centurion Way is another reference to the Roman road. If a theme can be considered anything more than one name, then Greek mythology can be said to have provided the names of Adonis Close, Iris Close and Flora Close. Mythology also gave names to Jason Close, Arion Close and Minerva Close. Oddly this theme was abandoned for the next road along is called Garage Close and then the next Redwell Close, which, oddly, relates to neither. However, nearby Jessop Drive does remember the former business premises here.

When any town is first developed, the existing field and other minor names are the natural choice for street names. Ostensibly this all took place in the Victorian era when farm workers and their families moved into the towns to seek the fortune and employment promised following the boom of the Industrial Revolution. There are some exceptions to this date, Ninefoot Lane being an old name describing itself. Cut on what is described as Park Field on a map of 1810, we find this reflected in the modern names of Parkfield Close, Parkfield Avenue and Parkfield Crescent. The lane known as Kinsall Green is the sole surviving indication of this former minor place name. Nearby is Relay Drive, a more subtle modern name than usual, which describes a road that gives access between the modern routes of the A5 and M42, an industrial estate and a service area.

The same map shows that Lowerpark also comes from a field name, that Bamford Street was land held by Richard Bamford, and Whateley Lane by Thomas Whateley. Perrycrofts Crescent is derived from the district name, itself taken from the four or possible five fields here named Perry Croft on the map of 1810. Other names derived from the same region are Croft Street, Perry Close and Scroft Avenue, while School Walk reminds us that this connects the college and the Queen Elizabeth's Mercian School. Riverfield Grove aptly describes its position close to the River Anker, while that of Overwoods Road is named after the 'higher woodland region'.

Hopleys Close was cut on the region covered by Hopleys Farm. Coton Lane shows a minor place name, as does that of Cottage Farm Road in Dosthill. Dunstall Lane is named after Dunstall Farm and is probably a former, and otherwise lost, minor place name. Gillway Lane cuts right through the region known as Gillway, this being a minor place name here for many years. Gorsy Bank Road was a minor trackway where gorse was the dominant shrub for many years before any idea of development was suggested, as was Hedging Lane and Old Hedging Lane. Greenhill Close and Orchard Close were likely the local names at Dosthill for years before the road was developed. Park Road is the remnant of an old winding road leading to what is now known as Dosthill Park. Parkside is a modern landscaping feature, and Park Street dates from several decades earlier.

Hanbury Road, Hanbury Court and even Hanbury's Farm Primary School could all point to the family of Robert Hanbury who once lived at Bole Hall. Kennedy Close is most likely associated with James Kennedy Esq. MD. In 1688 Ann Ferrers became the second wife of Robert Shirley, eldest son of Robert baron Ferrers of Chartley, and through a strange number of hereditary oddities (none of which were in any way suspicious) the castle and its lands fell to the one heiress, Elizabeth Shirley, who gave her name to Shirley Walk. Stanhope House remembers Philip Stanhope, Earl of Chesterfield, who was the second husband of widow Ann Ferrers in the early seventeenth century.

Cherry Street, Tamworth.

Hospital Street was the site of Tamworth General Hospital. Basin Lane leads to the Glascote Basin on the Coventry Canal, Bridge Street crosses it and Brindley Drive commemorates James Brindley who developed the technology to build canals. Guys Close remembers Thomas Guy, a benefactor of Tamworth who built the town hall and also gave the town its almshouses. Bowling Green Avenue is named because of the long-established bowling green here.

Spinning School Lane was named after the school built in the nineteenth century, where the modern police station now stands. Castle View does indeed afford a view of Tamworth Castle. Farm Close does adjoin farmland. Orchard Street at the Leys is cut in the region of the cherry orchard, clearly marked on a map as recently as the nineteenth century – also the origin of the name Cherry Street. Woodhouse Lane, which has also given its name to the school here, is an old landscape name.

The recently dug Borrowpit Lake has created a wetland habitat, albeit an artificial one. Roads nearby are named after plant life we would expect to find around such habitats and include Campion Drive, Reedmace and Tansy. Heron Court can also be considered an example. Brook Avenue and Brookside Way are named from the nearby Kettle Brook. Fontenaye Road is an odd name which is probably a reference to a family name. Indeed, the only other potential source for this name is a vessel known to have sailed to New Zealand and back on three occasions in the 1780s and 1880s. As the road was developed decades later, it is impossible to see a link between the two. New Road and the two New Streets are clearly all named such for they were indeed once 'new'. However, what might have seemed logical at the time is now clearly of little meaning.

The individuals who contributed to and influenced the development and prosperity of a town naturally gave their own names to newer developments. Beauchamp Road features the name of the family who were Earls of Warwick, and their device is shown on the nave of St Editha's Church, an indication of their link with Tamworth. The same church gave Church Lane, Church Street and Little Church Lane. College Lane was once known as Coket Street, after the family called Coket who had land here. The change in name refers to the College of Deans attached to the Church of St Editha's. St John Street takes its name from the dedication of the Catholic Church, here before the lane was developed. Church Road at Dosthill speaks of the Norman church here. Dedicated to St Paul, it was built in the twelfth century and, although no longer a place of worship, doubles as the venue for the Sunday school, parish meetings and the like. Camden Drive is named after Mr Camden, a historian of the town. Sharpe Street recalls mercer John Sharpe, who was trading in Tamworth by 1613.

Canning Road remembers Charles Canning, the owner of the colliery. Torc Avenue marks the site of one of the most valuable archaeological finds in all of south-east Staffordshire. Made somewhere between AD100 and 100BC, a gold-alloy torc of the Iron Age was found in a boatyard beside the Coventry Canal during the Second World War. Amazingly, it was a further thirty years before anyone realised exactly what had been uncovered that day or its true value. Such was the value of this find that it spawned a second street name here, that of Treasure Close. Goostry Close and Goostry Road are named after a family who lived here. William Goostre, together with his wife Agnes and son William, died at Hopwas on 27 January 1550. Never a rich man he left all to his family and his will named his (now married) daughter as beneficiary of 'God's blessing and mine but nothing else'.

Copes Drive may refer to Thomas Cope, a yeoman farmer of Wigginton, who left in his will of 5 February 1621, ten shillings (fifty pence) per year to ten poor people of Tamworth to be distributed on Good Friday and a further ten shillings (fifty pence) to ten poor people of Wigginton to be distributed on the Friday before Christmas Day. Or perhaps this is Charles Henry Cope Esq., a magistrate of Wigginton Park. Levett Road is named after the family that influenced this part of the county for many years, Thomas Levett of Croxall Hall being just one example. Tempest Street commemorates William Tempest, a magistrate who lived at Mayfield House. Bailey Avenue is thought to remember the will of Stephen Bayly who, having made his money in sheep farming, bequeathed an annuity of forty shillings in his will of 1 May 1620. Harcourt House remembers Michael Harcourt and Robert Harcourt, both of whom represented Tamworth in parliament.

Neville Street points us to Robert Nevill of Glascote, here in the nineteenth century, and Nevill Street (despite the different spelling) is undoubtedly of the same origin but a different generation. Argyle Avenue recalls the firm of solicitors who are still in business in the twenty-first century and were also in Tamworth in 1896 when Thomas Argyle was a legal man who was also clerk to the Borough of Tamworth. Crutchley Avenue is a reminder of the family who traded in the town at the end of the nineteenth century. Charles Crutchley was a provision merchant at 56 Church Street, while George Crutchley was a commerical traveller based at 5 Victoria Road. George Davis was a shopkeeper who also owned refreshment rooms in Bolebridge Street in 1896, and who is remembered by Davis Road.

Dent Street is named after Charles Dent, who was trading at the end of the nineteenth century from premises at 34 George Street. Dent and his workforce proved capable of a number of talents, advertising their services as 'painters, art decorators, plumbers, gas, steam, hot and cold water and electrical fitters, lead, glass, oil and color merchants'. Surely such diverse skills would have kept them in business through many a financial crisis.

Fenn Street remembers Charles Fenn, a colliery agent based in Tamworth in 1896. Thomas Heath & Son, boot and shoe dealers in the 1890s, were based on the corner of Colehill and Victoria Road and are the family who gave their name to Heath Street. Marshall Street is probably named after Charles Brownlow Marshall, who leased land for the mine sunk at Alvecote in the 1870s. David Morgan was a brewer's agent at Morgan & Co., Lady Bank, and it was he that gave a name to Morgan Road. There is also Shelton Street, which reminds us of Shelton Bros, a family of bakers at Smithy Lane, amongst whom was numbered Miss Mary Shelton, who kept a shop at 99 Lichfield Street.

Repington Road North and Repington Road South can be traced back to Thomas Repington, mentioned as the owner of the deanery of the Church of Tamworth in 1583. The Repingtons are a long line, originally from Lincolnshire, and one which continued in Tamworth for many generations. Thomas Guy Way is named after the benefactor who funded Tamworth Almshouses and Tamworth Town Hall, as well as the hospital, for which he is better known to the nation as a whole. MacGregor Crescent and MacGregor Tithe both refer to William MacGregor, a benefactor who was renowned for his modesty and tireless work, and whose real influence was only known after his death. Fox Close is probably named after Thomas Fox, who was owner of the Moat House by the time he married Judith Boothby on 28 September 1654. However, George Street, formerly known as Bull-Stake or Bullstoke Street, is not named from a person but from the George Inn which once stood there. On the 3 March 1716 Lady Elizabeth Shirley, heiress of the castle, married James Compton, 5th Earl of Northampton. Of their two sons and five daughters, the inheritance fell to the daughters Charlotte and Jayne Compton, Charlotte eventually marrying George Townshend in December 1751. The family name is remembered by Campton Road.

Townshend Close could refer to any or all of the family associated with Tamworth. Lord Charles Vere Ferrers Townshend (of Rainham, Norfolk and 23 Cavendish Square, London) and Captain John Townshend RN (of 53 Eaton Square, London) were both benefactors in Tamworth during the nineteenth century. There is also a Marquis Townshend, Earl of Leicester, who began life in Norfolk and who had a long association with the town, even considering to turn the Crypt of St Editha's into a family tomb! Wesley Way remembers the man who was a bookseller from Burton on Trent and who was associated with the town. Weymouth House is named after Lord Weymouth who gave a barn as part of what would become the workhouse. Willington Road is named after Francis Willington, who was Town Clerk during the nineteenth century.

There is evidence of the Florendine family being around Amington since the nineteenth century, and undoubtedly they gave their name to Florendine Street, which in turn gave its name to the school. Similarly Greenacre Close took the name of the Greenacre Primary School – no old field name but a created name. Hillcrest Close still stands at the summit of the hill, as does Hill Top Avenue. The Green at Amington perfectly describes the road which, although it now surrounds the church of St Editha's, certainly pre-dated it. Tolson Close at Dosthill remids us of the William Tolson who was a major employer at Fazeley.

Grazier Avenue remembers Tamworth's most famous hero of the Second World War, and the Colin Grazier Hotel is also named after him. Even when Bletchley Park started to decipher the encrypted Nazi messages, the U-Boat version of Enigma proved much more difficult. Not until 1943, after British sailors drowned trying to recover an enigma machine, were U-Boat codes broken on a regular basis. It was not an easy chase on 30 October 1942. HMS *Petard* and three other British destroyers were pursuing a Nazi sub in the Mediterranean Sea, not far from the Egyptian shore, and *U-559* was proving why U-boats were so dangerous. She was elusive, the chase lasted sixteen hours before *U-559*'s commander decided to scuttle his damaged ship about seventy miles north of the Nile Delta. Demonstrating courage that is hard to comprehend, three young men swam from *Petard* to *U-559*. Lt. Tony Fasson, able seaman Colin Grazier, and canteen assistant Tommy Brown descended into the sinking submarine. They wanted to retrieve the enigma machine. Also named to commemorate this truly heroic episode was Bletchley Drive, Fasson Close, Petard Close and Brown Avenue.

Not only local characters but also national figures and events have provided a wealth of potential names for streets. In November 1843 the Peels were visited by the monarch and her consort, the roads they took into the town from the station were renamed Victoria Road and Albert Road in their honour. A more recent development expanded upon this theme with Alexandra Mews, which one would guess refers to the queen, albeit that, correctly, it should be Alexandrina. Similarly Consort Place, off Albert Road, reminds us that Prince Albert was the monarch's consort. Brunel Close seems an odd choice in a region where wordsmiths are the theme, for it is clearly named after the famous engineer Isambard Kingdom Brunel who, among other things, is associated with the Great Western Railway and the modern Trent Valley Line that runs almost parallel to it.

Such individuals and events are not always restricted to historical developments. Anker View and Anker Drive are named after the river, while Ankermoor Court refers to the moorland through which the river flows to thc east of the town. Borough Road is taken from the region of Borough Park, itself named after its location on the northern limits of the Borough of Tamworth. Historically, the lower western part of Lichfield Street was known as Bradford Street, although today the name has been taken for a side road. The original was named for the 'broad ford' leading to the island and the broad meadows over the part of the river always known as *Wean-lake*. The owner at the end of the thirteenth century took the name of this region and is recorded as Robert de Bradford. Further references to the meadow are seen in nearby Meadow Park.

In the 1960s and 1970s Tamworth's population exploded. An agreement to provide housing for the overspill from neighbouring places, particularly Birmingham, meant huge areas were developed. Planners would have been stumped for names in the past, doubtless ending up drawing on friends, family and neighbours for inspiration. However, these wells soon run dry and, with the greatest respect to those who have been so honoured, some individuals have done little to merit such an honour.

Tamworth authorities made the decision to avoid naming roads after individuals, choices will always prove controversial with someone, and so a number of themes were created to provide a specific area with a link. The choice of theme is mostly random, although developers have had some influence, and some of the themes have been stretched to accommodate later additions. However, it does make such streets easier to locate, assuming you know the origin of the street name, and the newer streets have the benefit of being arranged in alphabetical order making directions much more simple. Furthermore these new estates have only a single name, with none of the usual street, road, avenue, drive, etc. which was in vogue at the time and which has since been thankfully avoided whenever possible.

One of the simplest of themes was chosen for the Amington Estate: that of woodland plants. This produced the street names of Abelia, a popular garden shrub from a family found across the world; Angelica, a group of sixty or so species of perennial herbs growing in the sub-arctic regions of the Northern Hemisphere; Briar, the common name given to several unrelated, but always thorny, plants; Brookweed, which refers to any of a dozen water-loving herbs; Clematis, mostly recognised as climbing plants, although some are officially shrubs, there are around three hundred species with hardly a garden in the land not having at least one example at some time or other; Cornel, a species of dogwood which, unlike most woods, is denser than water and does not float; Foxglove (there are about twenty species of woodland plant which are commonly referred to as foxgloves); Greenheart is not really a woodland plant, just a common name for several different trees in the Americas; Harebell (a plant similar to the bluebell but with variable colours from white, through pink and blue to violet); Hornbeam which refers to any of several medium-sized hardwood deciduous trees, listed as part of the birch family; Jasmine, a member of the olive family of trees, the majority are climbers and can occur as evergreens or deciduous varieties; Juniper which refers to any of over fifty varities of tree which occur naturally only in the northern hemisphere, and Juniper berries are used as a spice, notably in the flavouring of gin; Kerria is a deciduous shrub of the rose family native to eastern Asia; Lindera (any of eighty or more flowering plants native to Asia and commonly referred to as Spicebush or Benjamin Bush); Madrona is the common name of the genus *arbutus* and derived from the Spanish name for these species, it is native to the Northern Hemisphere; Magnolia, of which there are over 200 species, is an ancient genus dating from almost 100 million years ago, and is one of the earliest and most simple flowers lacking distinct sepals and petals; Maple Rise, correctly called the acer, with the distinctive pointed leaves common to most of the 120 or so species – its image is found on the national flag of Canada; Nemesia, a group of about sixty annuals, perennials and small shrubs most of which are cultivated hybrids from the native African stock; Quince, a small deciduous tree native of Asia and producing a fruit which is used to make preserves; Robina, a deciduous tree native to North America and named after Jean Robin who brought them to Europe in 1601; Saffron, which is not a woodland plant but a spice obtained from the dried stigma of a crocus, and which, by weight, is the most expensive spice known; Sorbus is a tree few people will have heard of in the UK, although they would possibly know it by its North American name of Mountain Ash and certainly know it by its UK name of Rowan; Sorrel, which refers to the Common or Garden Sorrel, a perennial herb cultivated as a leaf vegetable; Spruce is a coniferous evergreen cultivated in northern temperate zones in order to manufacture paper; Tilia Road, a large deciduous tree found across the Northern Hemisphere and better known in the country as the Lime; Trefoil, a name derived from the plant having a three-lobed leaf, with bird's foot trefoil being just on of over one hundred species around the world; and finally Woodland Road which is not only the arterial road through the estate but the key to the theme.

Motor cars provided further inspiration: Allard, the Allard Motor Company was founded by Sydney Allard in 1936 and produced 1,900 cars over the next thirty years;

Alvis, a Coventry-based manufacturer of military and luxury vehicles from 1919 to 1967; Bentley Way, Bentley Motors Ltd was founded by Walter Owen Bentley in 1919, later a part of Rolls Royce it has been owned by Volkswagen since 1998; Libra Close, released as a kit car in 1998, it is a lightweight sportscar; Belsize, a range of cars made in the early twentieth century; Cowley, a place in Oxfordshire where Morris cars were made and which gave its name to the Morris Cowley; Dennis, from Dennis Specialist Vehicles of Guildford who make buses, dustcarts, fire engines; Healey, from the Austin-Healey manufactured sports cars from the 1950s; Hillman, a famous name of motor car produced by the Rootes Group; Jaguar, from the famous quality cars produced in Coventry, where it was founded in 1922 by Sir William Lyons; Jensen, from the company based in West Bromwich, a manufacturer of sports cars and commercial vehicles; Jowett, after a company founded by William and Benjamin Jowett in Bradford and which was in business from 1906 to 1954; Lagonda, from a British manufacturer founded by American Wilbur Gunn in Staines, Middlesex; Leyland Road, named after the place in Lancashire which gave its name, most famously, to the British Leyland group; Lotus, a highly desirable range of sports cars from a manufacturer founded by Colin Chapman; Napier, a company which in the 1920s and 1930s Napier produced the most powerful motor car engines in the world; Riley, a motor car manufacturer who also produced bicycles when founded in 1890, the company became part of the Nuffield Organisation in 1938; Sunbeam, founded by John Marston in Wolverhampton in 1888 and producing bicycles, later the company produced cars until 1936, the Sunbeam was the first British car to win a Grand Prix race; Swift, a Coventry-based manufacturer producing vehicles from 1900 to 1931; Talbot, founded in 1903, a company which produced a range of vehicles until 1986; Triumph, based in Coventry the company made cars until 1984; Trojan, which produced cars from 1914 to 1974 and was based in Clapham; Wolseley, based in Birmingham, whose cars were produced under this name from 1901 to 1975; Witney, after the town in Oxfordshire which saw production of the Morris car range; and finally Wyvern, a range of cars produced by Vauxhall from 1948 to 1957.

Whilst it may seem identical, the theme for the following names is the Reliant Motor Company. Scimitar Close is named after the much lauded sports car produced under the Reliant name. Regal Close and Robin Close are also named after models, while Tom Williams Way marks the achievements of the company's founder. Nearby Quarry Hill, which itself gave a name to Quarry Court, refer to former workings here on the original route of the Watling Street, which is also the reason for Sandstone Court. Such a major road will have had several hundred blacksmiths along its route – what better place to find work – and one may even have been resident at Smithy Lane.

Industrial estates often show a suggestion of engineering or some other aspect of industry. Such names as Alloys Close, Amber Close, Amber Court, Bronze Court, Cupronickel Way, Felspar Road, Galena Road, Sandy Way, Pebble Close, Pewter Court and Silica Road refer to minerals, alloys and ores.

However, the Lichfield Industrial Estate chose the theme of space for its names, which produced an array of landmarks in the exploration of space by humanity. Anders, Borman and Lovell remember how William Anders, Frank Borman and Jim Lovell became the first people to leave the orbit of the Earth when they orbited the Moon aboard Apollo VIII in December 1968. Jim Lovell flew four missions: Gemini VII, Gemini XII, Apollo VIII, and the aborted Moon landing mission Apollo XIII. The latter missions gave a name to the road Apollo, whose most famous astronaut gave his name to Armstrong. Neil Armstrong (born 5 August 1930, Wapakoneta, Ohio) only flew two missions: Gemini VIII, when he piloted the first mission to achieve a docking in orbit; and Apollo XI, when he became the first man to set foot on another body anywhere in the universe when he walked on the moon.

Ariane is named after the three-stage rocket used by Western Europe to launch payloads into orbit. Gagarin marks the achievements of Yuri Gagarin, the Soviet cosmonaut who was the first man in space on 12 April 1961. He died in a plane crash in March 1968 aged thirty-four. Kepler reminds us of Johannes Kepler (1571-1630), the German astronomer who gave his name to the laws of planetary motion. Mariner was the name of the ten NASA missions to explore the inner rocky planets of Mercury, Venus and Mars. All fly-by missions, they were the first to photograph other worlds in the Solar System. While Neander is a lunar impact crater in the Sea of Nectar and was named in honour of Michael Neander, a German astronomer. Cavendish takes the name of the radio astronomy group based at Cambridge University. Gerard remembers Gerard Kuiper, an astronomer whose observations and theories on the Solar System were subsequently shown to be correct by improved technologies and probes. Langrange is named after Joseph Louis Lagrange (1736-1813), an Italian mathematician and astronomer who made great steps forward in the science of celestial mechanics Landsberg is named after a crater on the moon and the place where Ranger Seven impacted during the early days of space exploration.

Abbey Road leads to the region around Glascote named after abbeys and include Byland, Cleeve, Furness, Lindisfarne, Wenlock and Woburn. In Dosthill, as far south as any region covered by this book, road names reflect the nations' long love affair with horses; Alverton Close, Cadogan, Epsom Close, Felstead Close, Haydock Close, Kempton Drive, Larkspur, Morston, Sefton, Ascot Drive, Nimbus, Parbury and Rosebery Road. An Arabian stud, founded in 1896 when Colonel Lyon bought the mare Howa from the sales at Crabbet Park, gave its name to Harwood Drive. Lochsong Close names the bay thoroughbred, born in 1988, which won the Cartier Award as the top sprinter of the year twice and was voted 1993 European Horse of the Year. Arkle was the name of a superb National Hunt racehorse from Ireland who, during the 1960s, set standards which will probably never be matched. He won twenty-seven of his thirty-five starts, over distances from a one and a half mile sprint to almost four miles. Injury put a premature end to his career and, suffering from advanced arthritis, he was put down when he was only thirteen.

Country houses is a refreshingly new idea and has provided Braham, Browsholme, Buckingham Road, Chalfield, Chartwell, Chatsworth, Gawsworth, Glyndebourne, Godolphin, Helmingham, Kentwell, Littlecote, Longleat, Lorton, Melford, Newstead, Osbourne, Rufford, Swallowfield, Thoresby, and Wynyates. Royal residences produced the ideas for Gatcombe Place, Kensington Drive and Balmoral Close.

The university colleges of England have provided names for Blackfriars Close, Downing Drive, Exeter Drive, Greyfriars Drive, Lincoln Avenue, Mansfield Close, Oxbridge Way, Pembroke Close, Somerville Court, and Trinity Drive. While the names of Castle Road and Liberty Road could hardly be considered to be a theme, they do have a common origin in the old ward of Castle Liberty, today only seen in the name of the electoral ward.

Prime ministers are a common theme, albeit only referring to those of yesteryear in order to avoid any political conflict. Balfour recalls Henry Balfour; Palmerston Avenue,

Henry John Temple, 3rd Lord Palmerston (1784-1865), who served two terms as prime minister and held several other political offices during his career; Russell Close, John, 1st Earl Russell (1792-1878), a Whig politician who served two terms as prime minister and was also an accomplished writer and biographer; Peel Drive, Sir Robert Peel (1788-1850), who served two terms as prime minister and held several other political offices during a successful political career; and Peelers Way a reference to former Tamworth MP Sir Robert Peel who is credited with founding the modern police force and who were known as Peelers.

Rivers named Avill, Cherwell, Cringlebrook, Derwent, Hamble, Hayle, Honeybourne, Irwell, Kennett, Medina, Carey, Stour, Medway, Nymet, Tarrant, Wandsbeck, Waveney, Thurne, Seaton, Shannon, Ryton, Purbrook and Rothay gave their names to the streets in several places of Tamworth. Colbrook is also found here, but is erroneous for it should be the Colebrook, a popular fishing river in Northern Ireland, while Holsworth similarly should be Holdsworth. Redlake is not an actual river but a 'watery' name, the same is true of Romney which alludes to the Romney Marshes. Devon rivers produced Tamar Road, Teign, Torridge, Avon, Dart, and Ottery, while the local Tame was the basis for Tame Close and Tame Street. Bramber is a development among the 'river names' and yet there is no river of this name, which means we are probably looking at Bramber Bridge which crosses the River Adur in West Sussex.

A theme of railways is seen around Glascote Heath with Caledonian, a preservation railway in Scotland; Cambrian, a restored line near Oswestry; Caradoc, a design of 0-4-0 side tank locomotive; Carisbrooke, a location on the Isle of Wight railway; Chapelon, referring to Andre Chapelon, for thirty years a leading light in the design of French locomotives; Collet, a style of coach first seen in 1948; Deltic, the famous diesel locomotive which was seen as the ideal replacement for steam; Dunedin, the best known station of the wondrously scenic Taieri Gorge Railway in New Zealand; Faringdon, a Berkshire preservation line; Granville, when the estate was being developed the worst disaster in Australian railway history occurred near Sydney when eighty-three people lost their lives; Gresley, Sir Nigel Gresley is the best known engine designer in British railway history; Hawksworth, F. W. Hawksworth was a locomotive designer working at Swindon; Ivatt, remembers engineer and designer Henry Ivatt; Kirtley, named after Matthew Kirtley of the Midland Railway; Maitland, a historic line in New South Wales, Australia; Pullman Close, doubtless the most famous designs of railway coaches; Signal Walk, an obvious reference to the railway signalling system; and Stephenson Close, named after the man who started it all, George Stephenson. Not related is the name of Stationfields, which relates to the current railway station at Tamworth which stands alongside. All these run off Silver Link Road, which can only refer to the train service in and around London.

However, the oldest of all these names, and presumably the inspiration for the railway theme, is that of Engine Lane for all trains require some form of power. Yet the origins of this name were totally misunderstood (perhaps deliberately) by developers. Here the reference is to the coal mining in this area, indeed this is exactly where the winding engine for the mine once stood. Ingram Pit Lane is another reminder of one of Tamworth's now lost industries, and the surname has existed in the region for many years.

When first examining the names on the Tame Valley Trading Estate it seemed the theme was one of shipping forecast areas. However, two or three street names simply did not fit and a second search discovered the shipping forecast areas had been used to name

North Sea gas and oil field developments. From there came the names of Alwyn, which is 275 miles north-east of Aberdeen and has been operating since 1987; Brent, one of the first of the Shell oil fields, named after water-birds and standing in 460 ft of water, it is 116 miles north-east of Lerwick; Claymore, which has been in production since 1977; Forties, worked by BP from 1975 until 2003, is the largest of the North Sea oil fields; Magnus, found 100 miles north-east of the Shetland Islands, has been providing oil since 1983; Vanguard, an Australian company involved in the exploration of four fields; and Viking, one of the earliest oil fields to begin production, it is 86 miles of the Lincolnshire coast and has been worked since 1972.

Ninian Way, and from that Ninian Park, is named after the field which was apparently predicted by Scotland's answer to Nostradamus, the Seer of Brathan. The Ninian Central drilling platform was not the normal four-legged variety but a one-legged variation known as a mono-pod. Built at Kishom the burning flame of gas and single leg has been likened to the prediction of the appearance of a 'one-legged, fire-breathing giant from Nigg'. The authenticity of this prediction is slightly less credible than the existence of the man himself, known in his native tongue as *Coinneach Odhar* and suggested to have been a member of the Mackenzie clan from Uig. He is reputed to have made his predictions by gazing into a stone with a hole in the middle, where he would see the visions. Geologically speaking, all of these oil and gas fields are linked by a single sunken valley, hence the name of Valley Lane.

A theme of hills is pertinent to the high ground which marks the crest of the hill between Tamworth and Polesworth, indeed it is also the point where Staffordshire and Warwickshire now meet. This has produced Abberley – Abberley Hill is found in Worcestershire near the village of the same name; Amicombe – Amicombe Hill rises to 1,840 ft above sea level in the north of Dartmoor and is one of the bleakest, wildest and remotest regions on the moor making it worthy of its alternative name of The Devil's Dome; Bellingham – village and hill in Northumberland which is a favourite stop on the Pennine Way and, please note, should be pronounced 'Bell-in-jum'; Birdhope – actually known as Birdhope Craig is a very remote spot of the north-east Kielder Forest and is less than ten miles from the border with Scotland; Blackdown – the Blackdown Hills form the border between Somerset and Devon, and it has been declared an Area of Outstanding Natural Beauty and boasts sixteen Sites of Special Scientific Interest, each for its particular wildlife habitat; Brendon – five miles south of Watchet in Somerset at the extreme east of the Exmoor National Park is Brendon Hill, iron ore was extracted from mines opened at 1,000 ft above sea level until the twentieth century; Broadlee – Broadlee Hill stands a couple of miles north of Hawick in the Scotland; Buckden – correctly called Buckden Pike it rises to 2,303 ft above the Yorkshire Dales and is a popular climb for walkers with several marked routes to the top, although beware of the cloud which often blankets the upper slopes; Chiltern – the Chilterns run along a line of 75 miles from Oxfordshire to the Bedfordshire/Hertfordshire border; Cheviot – the Cheviots straddle the border of England and Scotland are rise to 2,674 ft and are home to the wild cattle at Chillingham Park which have been here since the thirteenth century and are classified as wild animals; Cleasby – rising to 1,677 ft in the North Yorkshire Dales; Correen – some 40 miles west of Aberdeen, rising to 1,598 ft; Crigdon – Crigdon Hill is 25 miles north-west of Newcastle-upon-Tyne and reaches 1,243 ft; Cromdale – Cromdale Hill rises to 1,348 ft some 20 miles east of Inverness; Darnbrook – reaches just over 2,000 ft above sea level in the Yorkshire Dales; Deepdale – a Lake District summit some 816 ft above sea level; Deerhill – a ridge of land running down the Mull of Kintyre; Dovestone – the name of both a hill and the reservoir which overlooks the village of Holmfirth, and thus has been seen by all as it formed the backdrop to many a moorland scene in the long-running BBC sitcom *Last of the Summer Wine*; and lastly Mendip Way – the Mendips in Somerset is a ridge of limestone rising to a maximum of 1,066 ft at Beacon Batch and a site of much interest because it is home to several endangered species of wildlife.

An unusual theme of RAF bases gave names to Odiham Close, after RAF Odiham in Hampshire; Lyneham Close, RAF Lyneham in Wiltshire; and Mildenhall, RAF Mildenhall in Suffolk, now home to the USAF. Metfield Close seems to be a misnomer, but it is probably named after RAF Milfield in Northumberland; Wellesbourne is named after RAF Wellesbourne Mountford in Warwickshire; Scampton Way refers to RAF Scampton in Lincolnshire, once home to the Red Arrows display team; Harwell Close comes from RAF Harwell in Oxfordshire; Benson View is named after RAF Benson is also in Oxfordshire; Lakenheath references Lincolnshire RAF Lakenheath, also occupied by the USAF; and finally Manston View, which is named after RAF Manston, now Kent International Airport. Yeovilton is named after the RAF Naval Base in Somerset. Helston Close is another erroneous name, for this is a Cornish place name while the base is near Hounslow in Essex. The base was actually named Heston and was where former prime minister Neville Chamberlain arrived in 1938. He had just had talks with one Adolf Hitler, and had proclaimed 'peace in our time', waving what turned out to be a worthless peace of paper.

All the streets off the Pennine Way, Stonydelph are named after places associated with this 268 mile long footpath of the same name, which leads from the foothills in Derbyshire to end just across the border in Scotland. The Stoneydelph estate was a major development and there are a large number of roads still to be named. However, aside from the theme other criteria had to be addressed. To be a fair reflection of the entire length of the path there should be nothing controversial, and thus there are very few actual parish names. Furthermore, the names are given in alphabetical order north to south, which makes finding the places much easier for all concerned. Such a massive undertaking with such stringent rules makes errors inevitable and yet these are few and only in terms of names coming from outside what is generally regarded as the Pennines. Hence we find Calder, Camhouses, Carlcroft, Craven, Crossfell, Crowden Road, Ealingham, Edale, Edenfield Place, Ellerbeck, Eringden, Fossdale Road, Garrigill, Gayle, Gofton, Goldsborough, Grassholme, Greenlee, Grindsbrook, Haltonlea, Hanlith, Hartleyburn, Hawkside, Hazelgarth, Hebden, Hesleden, Holwick, Lintly, Litton, Lothersdale, Loughshaw, Lowdham, Lowforce, Maldale, Malham Road, Marrick, Mellwaters, Melmerby, Middlesmoor, Milburn, Mayfield, Marsett, Mickleton, Mossdale, Montley, Mossdale, Murton, Pennymoor Road, Rainscar, Ravenstone, Raygill, Ribblesdale, Scammerton, Standege, Swindale, Sykesmoor, Tanhill, Torside, Tutehill, Wainrigg and Wessenden.

Across the A5 by-pass is Lakeland Drive, which tells us the names of Coniston, Ullswater, Windermere and Buttermere, all of which are derived from the Lake District. Unrelated, but still a theme of sorts, are the street names of Brambling and Curlew, less well-known birds of Britain. An unexpected theme crops up showing the different eras of British history in Roman Way, Norman Close, Hanover Court, Windsor Close, Lancaster Close and Mercia Close. Elsewhere we find further evidence of the long history of Tamworth in Mercian Way, Roman Court, Saxon Close, Saxon Drive and Saxon Mill Lane and from that Mill Lane.

Writers and Wordsmiths have given us the names of Arnold Close, Matthew Arnold (1822-88); Bronte Court, for Emily, Charlotte and Anne Brontë who only wrote six novels between them; Browning Close, after Robert Browning (1812-89); Burns Road, for Robert Burns the national poet of Scotland; Byron Road, which remembers George Gordon, 6th Baron Byron (1788-1824); Chaucer Close, after Geoffrey Chaucer (1342-1400); Chesterton Way, named for G. K. Chesterton (1874-1936); Coleridge Close, which remembers Samuel Taylor Coleridge (1772-1834); Dryden Road, after John Dryden (1631-1700); Eliot Close, for George Eliot (1819-80); Elizabeth Drive, after Elizabeth Barrett Browning (1806-61); Goldsmith Place, which remembers Oliver Goldsmith (1728-74); Keats Close, after poet John Keats (1795-1821); Kingsley Close, for Charles Kingsley (1819-75), professor of modern history at Cambridge (1860), Canon of Chester (1869), Canon of Westminster (1873) and Chaplain to the Queen (1873); Kipling Rise, which remembers Joseph Rudyard Kipling (1865-1936); Longfellow Walk, after US poet Henry Wadsworth Longfellow; Masefield Drive, for John Masefield (1878-1967); Milton Avenue, in memory of John Milton (1608-74); Shelley Road, after Percy Bysshe Shelley (1792-1822); Thackeray Drive, named for William Makepeace Thackeray (1811-63); and Wordsworth Aveune, which remembers William Wordsworth (1770-1850), poet laureate from 1843.

Scottish poet Thomas Campbell (1777-1844) gave a name to Campbell Close; Lawrence Court is named after novelist D. H. Lawrence (1885-1930); Shakespeare Close, named after William Shakespeare, considered the greatest exponent of English literature his *First Folio* (basically a collection of his greatest works) was not released until twenty years after his death; Spenser Close is named after Edmund Spenser (1552-99), an English poet best remembered for his allegorical, epic romance *The Faerie Queen*; Stevenson Road is for Robert Louis Stevenson (1850-94), Scottish novelist, essayist and poet who is best remembered for *Treasure Island* (1883); Tennyson Avenue is for Alfred Lord Tennyson (1809-92), a British poet best known for *The Charge of the Light Brigade* (1855), but critically acclaimed for *Ulysses* (1859); Keble Walk remembers poet John Keble (1792-1866); Thomas Hardy Court was named for Thomas Hardy (1840-1928) British novelist and poet, best known for his *Far From the Madding Crowd* (1874), *The Mayor of Casterbridge* (1886) and *Tess of the d'Urbervilles* (1891); and finally Bloomfield Way, the most plausible candidate for which is poet and playwright Robert Bloomfield (1766-1823), not among the best known of his profession but the only one who fits this theme.

A theme of the counties of England is applied to the roads leading off County Drive and giving Cornwall Avenue, Cumberland Drive, Devonshire Drive, Dorset Close, Durham Close, Hampshire Close, Hereford Way, Kent Avenue, Westmorland Close, Huntingdon Close, Norfolk Drive, Northumberland Close, Rutland Drive, Somerset Close, Suffolk Way and Surrey Drive. Scottish islands were the inspiration for Shetland Avenue, Skye Close, Arran Drive and Orkney Drive.

A water theme appears for the estate cut around Brindley Drive. The theme chosen was one of British bays and includes Bexhill Drive, Bracklesham Way, Brancaster Close, Caister, Christchurch, Durlston Close, Falmouth Drive, Filey, Newport, Padstow, Peterhead, Redcliff, Sandown, Seafield, Selker Drive, Skipness, Stonehaven, Yalland

Avenue, Wembury, Whitley Avenue and Whitesands Close. Unfortunately, Emberton Way became corrupted somewhere within the halls of administration. Here the place name refers to Emberton in Buckinghamshire, which is many miles from the sea, but the intention was probably Embleton in Northumberland.

That old favourite theme of trees was the basis for Beech Close, Beech Road, Cedar Drive, Cherry Tree Walk, Chestnut Avenue, Hawthorne Avenue, Holly Avenue, Laburnum Avenue and Lilac Road. After trees birds are probably the second most popular theme and here are represented by Falcon Close, Goldcrest, Hawfinch, Kestrel, Kingfisher, Lapwing, Merganser, Merlin Close, Nightingale, Nuthatch, Osprey, Redwing and Sandpiper.

Straddling the border with neighbouring Warwickshire is Tamworth Municipal Golf Course. The roads in and around here have been named to complement the sport and include some of the best known courses in the land in Carnoustie, Gleaneagles, Hoylake, Lytham, Muirfield, St Andrews, Sunningdale, Troon and Turnberry. Eagle Drive features two golfing terms, one used to describe a hole completed in two shots under par and the second the long shot up to the green, while Fairway Court speaks of the close-cut grass on the main body of the hole.

The fairly common theme of water has been given an unusual and refreshing slant with names such as Boulder Close, Delta Close, Lagoon Road, Ripple Close, Rivulet Close, Torrent Close, Valley Drive, Waterfall Close and Meander Close. Fish is the theme on the new Two Gates estate where we find Bream, Chub, Dace, Grayling, Lamprey and Roach.

There is also an assortment of marine fish which are less likely to be known. Cobia is a large tropical marine game fish that can reach over two metres in length. Dorado, or *Coryphaena hippurus*, is a brilliantly coloured large marine fish known by the local fishermen as the dolphin. Gurnard is after the marine fish which got its name from the Latin *grundire* 'to grunt', because it grunts when it is caught. Houting is a European food fish, *Coregonus oxyrhynchus*, which lives in the sea but returns to breed in freshwater rivers and lakes. Johndory is named after the spiny-finned fish of the Atlantic and Mediterranean. The street name has been reproduced as one word, while the fish is strictly speaking two words.

Kurtus is named after an unusual fish, found off the shores of Australia and New Guinea where the males of two species carry the egg cases of the female on hooks protruding from the forehead for protection during incubation, as a result it is better known as the Nurseryfish. Manta Road is a large marine fish named from the Latin *manta* meaning 'cape', describing the shape of these large animals with a flattened shape. Marlin is a prized gamefish of Atlantic and Pacific waters, itself named from the marlinspike, a pointed metal spike used by sailors in parting the strands to splice two pieces of rope together and said to resemble the snout of the fish. Whiting is a North Atlantic fish, often seen on supermarket shelves.

A theme of British prime ministers is seen with Chatham Way, named for the Earl of Chatham, better known as William Pitt the Elder (1708-78). Considering his reputation it comes as a surprise to find he only served two years as what was effectively, if not officially, prime minister. St Christopher's Drive has been selected as he is the patron saint of travellers and the drive leads to the mobile home park, but, ironically, the mobile homes here are static and will never move.

A theme of castles is found at Two Gates for the names of Arundel, Bamburgh, Belvoir, Chillingham, Dunster, Morpeth, Slingsby, Sudeley and Tutbury. While the latest development in Tamworth is off the south end Marlborough Way where we find Alvie Walk, Garton Close, Gowan Close, Harport Close, Harrow Close, Lindor Road, Lyon Drive, Laggan Close, Leven Road, Lowes Drive, Maree Walk and Morton Close all feature a theme of Scottish lochs, bays and places with similarly water-related names.

Developers' names, kept as street names, include The Hedgerows and Lyndale, which actually tells us the developer was Lyndale Homes. Likewise, the most likely source of Galliers Close is Frank Galliers Homes. Rosewood Court is derived from Rosewood Close, itself a developer's name. Croft Avenue is named after Croft Street off which it is cut, and it is almost certainly an old landscape name.

Of course, there are always names which were created on the spur of the moment by a developer, planner or otherwise unknown individual who, even if they could be consulted, would probably be unaware of why that name was chosen. Examples in Tamworth include Blenheim Close, Belvedere Close, Allensmead, Broadsmeath, Arbor Close, Arden Close, Ash Grove, Ashdale Road, Ashlands Close, Barnbridge, Beechwood Crescent, Coach House Rise, Cottage Walk, Ivyhouse Walk, Leisure Walk, Linthouse Walk, Stonehill Walk, Bakers Walk, Callis Walk, Crestwood, Cygnet Drive, Ridgewood Rise, Highfield Avenue, The Dell, Elm Tree Walk, Fairview Close, Fir Tree Close, Grange Close, Greatmead, Kenilworth Road, Warwick Road, Langdale Court, Laurel Bank, Lavender Road, Linden Close, Lomita Crescent, Longlands Drive, Maxstoke Close, Meadow Street, Meadowwyrthe, Mount Pleasant, Oak Tree Walk, Orchard Street, Paddock Close, Pegasus Walk, Pine Close, Redhill Close, Rosemary Road, Rosewood Close, Rosy Cross, St Austell Close, St Ives Close, St Margaret's Road, St Peter's Close, Sherbrook Avenue, Skidmore Avenue, Solway Close, Stephanie Grove, Sunset Close, Swanmote, Sycamore, Tavistock Close, Tudor Crescent, Willowbank, Windmill Close, Woodcroft Avenue and Woodhurst Close.

Clifford Close is cut off and named after Clifford Street, itself one of several street names almost certainly named after individuals, although it is difficult to say exactly whom. Other examples of this are seen in Aitken Close, Alfred Street, Arkall Close, Ashleigh Drive, Aspbury Court, Barbara Street, Barcliffe Avenue, Barlow Close, Beyer Close, Bright Crescent, Browns Lane, Cale Close, Carlton Crescent, Chandlers Drive, Claremont Road, Clifton Avenue, Colbourne Road, Dormer Avenue, Grayston Avenue, Hampton Close, Hardwick Court, Hayworth Close, Heathcote Close, Henley Close, Hilmore Way, Jervis Road, John Street, Jonkel Avenue, June Crescent, Kilbye Close, Kimberley, Kingston Close, Leedham Avenue, Longfield Close, Madox Close, Marlow Road, Milo Crescent, Monks Way, Norton Close, Parson Street, Penny Court, Portland Avenue, Primley Avenue, Priory Close, Rene Road, Richmond Close, Robert Close, Scott Road, Steere Avenue, Stretton Street, Summerfield Road, Sutton Avenue, Thornby Avenue, Thornton Way, Tolman Drive, Welford Road, West Street, Wigford Road, Willoughby Road and Yenton Close.

There are also those street names where, while the meaning is fairly obvious, just why they were so named is unknown. Not knowing why, when or by whom any name was created makes defining it difficult; for example, Beech Avenue could refer to the tree or to an individual. Anchor Close would seem to share a common origin with the Anchor public house, that is both refer to the traffic along the canal which anchored here.

Arden Road probably refers to the forest that is still famous, thanks to the writings of Shakespeare, but it could also be a family name. Belmont Road brings to mind the famous American horserace, although just why this would be relevant to Tamworth is obscure.

The Gatehouse, Tamworth Castle.

Birds Bush Road sounds like a public house name, but it could be an old landscape name. Blythe Street is more likely to refer to the river name than to a family name. Today Common Lane is only a narrow and steep entry to the modern shopping centre, yet historically it was the route to the common where the locals could graze livestock.

Coronation Street was renamed to mark the ceremonial crowning of a new monarch at the beginning of the twentieth century, thus either Edward VII or George V. The naming of Corporation Street marked the early days of the development by the town council. Draycott Crescent could be a modern developer's name or may have been chosen to fit with nearby Drayton Bassett and Drayton Manor. Sheepcote Lane has either come from an old landscape name or possibly created to appear as such. Stonepit, too, takes its name from an earlier feature here.

Silver Street and Ladybridge Street were once known collectively as the Holloway. This major route out of Tamworth was in use for so many years that the horses and carts had literally worn a 'hollow way'. The name of Silver Street probably refers to the Tamworth Mint which was here for many years. The Riddings is an old term which is found as the street name at Amington. Today it is a cul-de-sac, passage of traffic stopped by the Coventry Canal although the bridge does allow pedestrians to cross. Ironically the original meaning speaks of it being where 'the trees were cut for passage' and from Middle English *ryding*.

Other names have only one possible origin, although why they were chosen is unknown and seems quite abitrary. Albion Street features the Latin name for England. Bancroft does not follow the theme for the rest of the roads here and is most likely an earlier field name or may be a remnant of the coal-mining era. Bridgewater Street could well refer to one of the places so named, although the surname theory seems more plausible. Similarly Burton Close is easier to see as honouring a family name than to have commemorated any of the many place names of England. It is possible that Castle Close would offer views of the castle, were it not for the obstructions, although it may also be a personal name. While Castlehall is a part of the area where the theme is the railways, it seems to have no link to that theme and as such may well be an error. Exley is another candidate for a potential error, it is found with other streets with a theme of rivers and yet there is no significant river of this name.

Launceston Close is the only one in this area not named after a river, indeed there is no River Launceston, the Cornwall town stands on the River Kensey two miles from the River Tamar. Clearly the thinking here was quite separate from the surrounding area and the reason for this name is unknown. Similarly Lomond Close in Leyfields has no link to other street names here. It can only have come from the famous Scottish loch, the largest lake of mainland Britain by surface area and which contains the island of Inchmurrin, itself the largest island surrounded by freshwater in Britain. The only other candidate would be the Irish racehorse which won the 2,000 Guineas in 1983, although again there is no known link to either explanation.

Marlborough Way forms the spine of the road through Belgrave and Glascote Heath. Two possible origins for this name, neither of which have any link to Tamworth, are the ancient market town of Wiltshire, or perhaps the Dukes of Marlborough who reside at Blenheim Palace and whose best-known family member was Sir Winston Churchill. Telford Road could be named after the Shropshire new town, or perhaps Thomas Telford after whom the town was named. Born August 1757, this Scotsman is remembered as a stonemason, architect, civil engineer, and builder of roads, bridges and canals and made many landmarks over the seventy-seven years of his life.

Prospect Street is a common street name found in many towns, the origins of which depend upon the period it was cut and its location. Earlier use invariably refers to the location, even today estate agents speak of 'good prospects' when they really mean a good view from the window. Later the term is used when a region is redeveloped, usually for industry, and thus improving the prospects of a region. The age of Prospect Street would suggest the former definition, although in the modern world the latter also fits the bill with the nearby Lichfield Industrial Estate. Queensway is a road name which always refers to some aspect of Queen Elizabeth II's reign, be it her accession, coronation, jubilee or a royal visit.

Dog Lane must surely refer to its shape, for there is a decided 'dog leg' in the lane, although a loyal pet is an alternative. Neighbouring Hodge Lane is almost certainly a family name, but could be a greatly corrupted (and thus unknown) minor place name. East View either comes from its location respective to the other roads of this development, its location on the east of the town, or is another family name. Springfield Road sounds like a field name, although records are unclear if this was the actual name of the place. Highcliffe Road probably features an old name in the landscape for it certainly fits, although it may also have been created to fit the topography. Next to it is Lansdowne Crescent, which may refer to the former Lansdowne Road stadium in Ireland which was demolished in 2007 to make way for a new national stadium; or it could be taken from the title derived from the place, the Marquis of Lansdowne would certainly have been around at the time of the Peel government. However, neither seem particularly strong candidates for naming this residential street.

In Amington, St Mary's Way and St George's Way are clearly connected by more than a junction, but that connection is not directly religious. Likewise, School Street no longer marks the site of a school. Tinkers Green Road would automatically be considered to be where tinkers had their home, yet it is now often said to be a derogatory term and has fell out of use in recent years. Historically it referred to a nomad tinsmith, who made and sold his own pots and pans to earn a living. The name of the street, however, could just as easily have come from a person with that surname or even nickname.

The name of Town Wall clearly does not refer to any defensive feature of the Tamworth town, for it is around two miles from the town centre. Here, we are probably looking at a corruption of an earlier name, now lost or never recorded. With its proximity to the original line of the Roman road of Watling Street it could be an old map record, although it seems unlikely if it will ever be really understood.

Falna Crescent must surely be named after the Indian town, which is one of the most important stops on the Ahmedabad-Jaipur railway line. It can also lay claim to being one of the largest producers of umbrellas on the globe. However, neither seem to have any connection with Falna Crescent and the link is probably an individual, and that connection was not recorded.

Halford Street was named later than Halford Terrace, the row of tenement houses (now gone) which were home to a cottage industry in the early and mid-nineteenth century. It seems the occupants of the houses probably worked together in the production of cloth or clothing and the name Halford was that of the owner or driving force behind the business. In the 1990s an elderly gentleman walked into the local Tourist Information offices. He was making a return to Tamworth for the first time since he was evacuated to the town at the age of six during the Second World War and was surprised to find high-rise flats where he had always known a textile factory. These are the prominent blocks off Lichfield Street and may have been later generations of a business, once in what is now Halford Street.

# Chapter Fourteen

# Wall—Woodhouses

## Wall

As the main arterial road found time and again throughout this book, the name of Watling Street is once again a reference to the tribe inhabiting the area around modern St Albans in Hertfordshire and tells us it was the road leading there. The only other destination name here is the one leading to the village itself and is Wall Lane.

Ashcroft Lane and Clay Pit Lane are both reminders of the landscape here. The Butts is another field name. It is often assumed to be a field where archery was practiced, the 'butts' being the targets, but it is more likely to refer to the strip of land left at the edge of the field where the plough is turned and is not cultivated. Green Lane, as with this name almost everywhere, was overgrown, and Market Lane still enables motorists to travel to nearby Lichfield market.

Roman Walk leads past the remains of the Roman settlement, which shows how important this staging post was on the Roman Watling Street. Indeed, a contemporary record of the day lists this place alongside such major cities as London, Lincoln, Chester, York and Wroxeter. However, this was more likely due to a lack of nearby staging posts on the major trunk road, making this an important overnight stop.

## Weeford

There are only three lanes in this hamlet south of Lichfield. All have etymologies from history and all are typical of Middle English names. Dog Lane is descriptive in speaking of an ostensibly straight lane except for a small 'dog leg' halfway down. Hungry Lane is often applied to an area which brought little return, more often than not because of the farmer's incompetence rather than the poor quality of the soil although which is the origin here is unknown. Flats Lane refers not to the land alongside being particularly even but to it lacking undulations, thus making it easier to work.

Here the church of St Mary was rebuilt in 1802 with the painted glass of the east window brought from Orleans in France by Sir Robert Lawley. It depicts Pontius Pilate on the judgement seat and washing.

The road marks the route, the signpost the destination.

## Whittington

Destination names for here include Whittington Lane, Whitting Common Road, Darnford Lane, Fisherwick Road, Huddlesford Lane and Burton Road (referring to Burton-on-Trent). Fisherwick Close is derived from Fisherwick Road, and Main Street was always the principal route through the village. Noddington Lane is the sole surviving reminder of a minor place name near Fisherwick.

Church Street is refers to the parish church dedicated to St Giles, Chapel Lane to the chapel which still stands there. Blacksmith Lane is, somewhat predictably, where the forge was located. The Green is still the traditional centre of the village, Back Lane is named after a field known as Back Croft and Marsh Lane is exactly what it would seem. Bramley Way is cut almost exactly where the orchard was planted, Spring Lane marks the site of an ancient water source and Rock Farm Road still leads to Rock Farm, although there were once two of this name.

Langton Crescent was named after Herbert Langton, who was both wheelright and village undertaker and whose forge was located here. The house and gardens of the Marquis of Anglesey provided employment for many and those employees have given their names to the streets here: Joseph and George Neale are remembered by Neale Croft, while Pass Avenue is named after the family which included Nathaniel Pass, John Pass, Charles Pass and William Pass. Dyott Avenue was land owned by Richard and William Dyott, Babington Close is named after the house built in 1673 by Zachary Babington and the pillars here are all that are left of the gate to this house. Samuel Seckham bought Whittington Old Hall and developed Cloister Lodge, and Cloister Walk is the somewhat circuitous route through the grounds.

These impressive pillars once formed an entrance to Fisherwick Hall.

Modern roads named after raptors include Falcon Drive, Kestrel Close and Merlin Way. Swan Road is not actually named after the bird itself but is named after the pub here. Darby Avenue remembers one of the great characters of twentieth century Whittington. Nurse Darby rode into the village on her bicycle in 1926. Throwing her bags into a room at Yew Tree Cottage for her stay of just one month, she looked forward to her temporary position attending to the medical requirements of the villagers. Forty years later in the 1960s she was at Buckingham Palace receiving the MBE from the queen for her service as district nurse and midwife.

During the eighteenth century, Whittington Heath was home to the Lichfield races, a very well attended annual event, which became a social event. In 1773 a grandstand was constructed offering fine views over the races, at the same time retaining the lifestyle the upper classes demanded. By the nineteenth century the event had lost its popularity and, with the sale of the land to the War Department, the Staffordshire Regiment moved in and the racegoers never returned. However, the grandstand is still there, now forming the base of the golf course.

## Wigginton

Comberford Lane is a destination name as is Syerscote Lane, although this is not the easiest to reach via a gated single-track road. Main Street has always been the primary road through the village. Portway is a common name, always referring to the 'way to market'. Willow Bottom Lane fittingly runs to the low wetland favoured by willow trees, while Green Lane is a typical name given to an overgrown country lane.

Commemorative plaque to the Babingtons of Curborough in St Chad's Church.

## Wimblebury

Brickworks Road marks where there was a brickworks in the nineteenth century, and Keys Park Road is an old estate name. Arthur Street may well refer to landowner Henry Arthur, while Dugdale Close, Glover Street, John Street, Knighton Road, Paul's Road and Scott Street all have names of familes which have been around here since at least the seventeenth century. This is not to suggest this is the origin but only a possibility, for records are unclear.

A theme of Australian place names is found for the names of Fremantle Drive – the city founded in 1829 by the English governor James Stirling, who named it after Captain Sir Charles Fremantle who had surveyed the area before him; Adelaide Drive, founded in 1836, was named in honour of Queen Adelaide, consort of William IV who was on the throne at the time; Brisbane Way – the city named in 1824 after its founder, Sir Thomas Brisbane; and Albany Drive, named by Governor George Stirling in 1832 after Prince Frederick, Duke of York and Albany and son of George III.

It is the norm for modern street names to be taken from a selection which suggest a rural location to potential buyers even before the property has been seen. In the past Acacia Avenue was seen as a fictional name, one which conjured up an image of suburbia. However, in recent times it has become the quintessential street name, one which instantly tells prospective residents what they should expect. Similar names here, created to fit the image, include Appledore Close, Barn Way, Beacon Way, Hawthorn Road, Haymaker Way and Kingscroft. Horseshoe Drive also fits into this idea, although it is not particularly accurate, describing the shape we would expect to find but do not. The idea of rural names has even developed into a theme in the area where Watersmead Close, Watermint Close and Waterlily Close were cut.

## Woodhouses

Grange Lane speaks of the country houses with farm buildings, probably those of Pipe Grange Farm. Green Lane refers to the common here. The Roche is from the Old English word for rock or stone, named for the stone quarried here. Camsey Lane shares its name with an old well on a map of 1845. This must be an old place name from a much earlier time.

# APPENDIX

# Suggestions

Traditionally street names reflect the history of the area. In the days before town-planning meetings the names developed naturally, coined by residents to reflect the area and those who lived there. As the population shifted from being mainly agricultural to being concentrated in towns and cities, people needed somewhere to live. And this led to house building on a scale never seen previously. Houses were lined up in streets, and the streets needed names. Thus, it was only right that those who had funded the building of streets, generally for their workers, would give them their own names. Soon family names were exhausted and other sources were sought. By the time local authorities took over the reins of the name game, the streets took on the name of the fields and those who had worked them prior to development.

Today, developers have taken to naming streets after themes, particularly themes which suggest a more rural location. This has resulted in towns and cities, and even villages, having networks of streets which mirror every other town and city and have no unique qualities whatsoever. Hopefully, this is merely a passing fad and soon we will return to more imaginative and (more importantly) relevant names.

Clearly there are any number of potentially controversial names which should be avoided, while the Post Office and emergency services require unique names which do not cause confusion. Themes are certainly an excellent idea, yet do we really need any more trees, birds and animals? Real place names are an excellent idea, yet surely those chosen should not be purely arbitrary but have some relevance?

Today, communication and rapid response times have meant many of these and other services can be provided from afar, but those members of the community of yesteryear are still remembered by some and would make excellent examples for street names. In the twenty-first century the public can nominate potential recipients of an award from Her Majesty's Government. With so many receiving the OBE, MBE or CBE on merit is this not the ideal source of street names in a community? The honour is soon forgotten yet the street name will likely last for decades at least.

In times past the employers, small businesses, the shopkeepers, teachers, bus drivers, clergymen, milkmen, postmen, undertakers, postmistresses, caretakers, policemen, doctors, midwives and nurses were pivotal to the community. People still organise neighbourhood watch, best-kept village entries, fêtes, fun runs, charity events, local football, cricket, rugby, tennis, athletics, boxing, darts, snooker, angling, youth groups, dance schools, the WI, church groups, coffee mornings, history groups – a multitude of subjects which, of course, also includes local authors.

# Bibliography

*A History of Hednesford & Surrounding Villages*, Anthony Hunt, Mount Chase Press (2005)

*A Walk Through the History of Whittington*, Whittington and District History Society, Lichfield Press (2003)

*A History of the Parish of Fazeley*, Tony Brookes (2000)

*Best of Cannock Chase* by Pitman, Midland News Association (1933)

*The Economic Development of the Marquess of Anglesey's Estate 1842-91*, D.G. Brown

*Probate Inventories of Tamworth*, Staffs County Council

*Tamworth Borough Records*, H. Wood (1952)

*Kelly's Directory of Staffordshire 1896 Part 1*

# Further Reading

*Hampshire Place Names*
Anthony Poulton-Smith

ISBN: 978-1-84868-725-7
Price: £12.99